Tucholsky Wagner Zola Scott Sydow Freud Schlegel
Turgenev Wallace Fonatne Twain Walther von der Vogelweide Fouqué Friedrich II. von Preußen
Weber Freiligrath Frey
Fechner Fichte Weiße Rose von Fallersleben Kant Ernst Frommel
Richthofen
Hölderlin
Engels Fielding Eichendorff Tacitus Dumas
Fehrs Faber Flaubert
Eliasberg Ebner Eschenbach
Feuerbach Maximilian I. von Habsburg Fock Eliot Zweig
Ewald Vergil
Goethe London
Elisabeth von Österreich
Mendelssohn Balzac Shakespeare Dostojewski Ganghofer
Lichtenberg Rathenau Doyle Gjellerup
Trackl Stevenson Hambruch
Mommsen Tolstoi Lenz Droste-Hülshoff
Thoma Hanrieder
von Arnim Hägele Hauff Humboldt
Dach Verne Hagen Hauptmann Gautier
Reuter Rousseau Baudelaire
Karrillon Garschin Defoe Hebbel
Damaschke Descartes
Hegel Kussmaul Herder
Wolfram von Eschenbach Schopenhauer Rilke George
Bronner Darwin Dickens Grimm Jerome
Melville Bebel Proust
Campe Horváth Aristoteles Voltaire Federer
Bismarck Vigny Barlach Heine Herodot
Gengenbach
Storm Casanova Tersteegen Grillparzer Georgy
Chamberlain Lessing Gilm Gryphius
Brentano Langbein Lafontaine
Strachwitz Claudius Schiller Kralik Iffland Sokrates
Schilling
Katharina II. von Rußland Bellamy
Gerstäcker Raabe Gibbon Tschechow
Löns Hesse Hoffmann Gogol Wilde Vulpius
Gleim
Luther Heym Hofmannsthal Morgenstern
Roth Klee Hölty Goedicke
Heyse Klopstock Kleist
Luxemburg Puschkin Homer Mörike Musil
Machiavelli La Roche Horaz
Navarra Aurel Musset Kierkegaard Kraft Kraus
Lamprecht Kind Kirchhoff Hugo Moltke
Nestroy Marie de France
Laotse Ipsen Liebknecht
Nietzsche Nansen Ringelnatz
Marx Lassalle Gorki Klett Leibniz
von Ossietzky May vom Stein Lawrence Irving
Petalozzi Knigge
Platon Kafka
Sachs Pückler Michelangelo Kock
Poe Liebermann Korolenko
de Sade Praetorius Mistral Zetkin

The publishing house tredition has created the series **TREDITION CLASSICS**. It contains classical literature works from over two thousand years. Most of these titles have been out of print and off the bookstore shelves for decades.

The book series is intended to preserve the cultural legacy and to promote the timeless works of classical literature. As a reader of a **TREDITION CLASSICS** book, the reader supports the mission to save many of the amazing works of world literature from oblivion.

The symbol of **TREDITION CLASSICS** is Johannes Gutenberg (1400 – 1468), the inventor of movable type printing.

With the series, tredition intends to make thousands of international literature classics available in printed format again – worldwide.

All books are available at book retailers worldwide in paperback and in hardcover. For more information please visit: www.tredition.com

tredition was established in 2006 by Sandra Latusseck and Soenke Schulz. Based in Hamburg, Germany, tredition offers publishing solutions to authors and publishing houses, combined with worldwide distribution of printed and digital book content. tredition is uniquely positioned to enable authors and publishing houses to create books on their own terms and without conventional manufacturing risks.

For more information please visit: www.tredition.com

Lameness of the Horse Veterinary Practitioners' Series, No. 1

John Victor Lacroix

Imprint

This book is part of the TREDITION CLASSICS series.

Author: John Victor Lacroix
Cover design: toepferschumann, Berlin (Germany)

Publisher: tredition GmbH, Hamburg (Germany)
ISBN: 978-3-8491-5542-1

www.tredition.com
www.tredition.de

PREFACE

All that can be known on the subject of lameness, is founded on a knowledge of anatomy and of the physiology of locomotion. Without such knowledge, no one can master the principles of the diagnosis of lameness. However, it must be assumed that the readers are informed on these subjects, as it is impossible to include this fundamental instruction in a work so brief as this one.

The technic of certain operative or corrective procedures, has been described at length only where such methods are not generally employed. Where there is no departure from the usual methods, treatment that is essentially within the domain of surgery or practice is not given in specific detail.

Realizing the need for a treatise in the English language dealing with diagnosis and treatment of lameness, the author undertook the preparation of this manuscript. That the difficulties of depicting by means of word-pictures, the symptoms evinced in baffling cases of lameness, presented themselves in due course of writing, it is needless to say.

It is hoped that this volume will serve its readers to the end that the handling of cases of lameness will become a more satisfactory and successful part of their work; that both the practitioner and his clients may profit thereby; and last but by no means least, that the horse, which has given such incalculable service to mankind and is deserving of a more concrete reward, will be benefited by the application of the principles herein outlined.

In addition to the consultation of standard works bearing on various phases of the subject of lameness, the author wishes to thankfully acknowledge helpful advice and assistance received from the publisher, Dr. D.M. Campbell; to appreciatively credit Drs. L.A. Merillat, A. Trickett and F.F. Brown for valuable suggestions given from time to time. Particular acknowledgment is made to Dr. Septimus Sisson, author, and W.B. Saunders & Co., publishers of The Anatomy of Domestic Animals, for permission to use a number of illustrations from that work.

J.V.L.

Chicago, Illinois, October, 1916.

Justice shows a triumphant face at the works of humane practitioners, who give serious thought and expend honest effort, for the alleviation of animal suffering.

6

TABLE OF CONTENTS

SECTION III
Lameness in the Fore Leg

SECTION IV
Lameness in the Hind Leg

ILLUSTRATIONS

INTRODUCTION

Lameness is a symptom of an ailment or affection and is not to be considered in itself as an anomalous condition. It is the manifestation of a structural or functional disorder of some part of the locomotory apparatus, characterized by a limping or halting gait. Therefore, any affection causing a sensation and sign of pain which is increased by the bearing of weight upon the affected member, or by the moving of such a distressed part, results in an irregularity in locomotion, which is known as lameness or claudication. A halting gait may also be produced by the abnormal development of a member, or by the shortening of the leg occasioned by the loss of a shoe.

For descriptive purposes lameness may be classified as *true* and *false*. *True lameness* is such as is occasioned by structural or functional defects of some part of the apparatus of locomotion, such as would be caused by spavin, ring-bone, or tendinitis. *False lameness* is an impediment in the gait not caused by structural or functional disturbances, but is brought on by conditions such as may result from the too rapid driving of an unbridle-wise colt over an irregular road surface, or by urging a horse to trot at a pace exceeding the normal gait of the animal's capacity, causing it to "crow-hop" or to lose balance in the stride. The latter manifestation might, to the inexperienced eye, simulate *true lameness* of the hind legs, but in reality, is merely the result of the animal having been forced to assume an abnormal pace and a lack of balance in locomotion is the consequence.

The degree of lameness, though variable in different instances, is in most cases proportionate to the causative factor, and this fact serves as a helpful indicator in the matter of establishing a diagnosis and giving the prognosis, especially in cases of somewhat unusual character. An animal may be slightly lame and the exhibition of lameness be such as to render the cause bafflingly obscure. Cases of this nature are sometimes quite difficult to classify and in occasional instances a positive diagnosis is impossible. Subjects of this kind may not be sufficiently inconvenienced to warrant their being taken out of service, yet a lame horse, no matter how slightly affected, should not be continued in service unless it can be positively established that the degree of discomfort occasioned by the claudication is small and the work to be done by the animal, of the sort that will not aggravate the condition.

Subjects that are very lame — so lame that little weight is borne by the affected member — are, of course, unfit for service and as a rule are not difficult of diagnosis. For instance, a fracture of the second phalanx would cause much more lameness than an injury to the lateral ligament of the coronary joint wherein there had occurred only a slight sprain, and though crepitation is not recognized, the diagnostician is not justified in excluding the possibility of fracture, if the lameness seems disproportionate to the apparent first cause.

The course taken by cases of lameness is as variable as the degree of its manifestation, and no one can definitely predict the duration of any given cause of claudication.

Because of the fact that horses are not often good self-nurses at best, and that it is difficult to enforce proper care for the parts affected, one can not wisely state that resolution will promptly follow in an acute involvement, nor can he predict that the case will or will not become chronic. Experience has proved that complete or partial recovery may result, or again, that no change may occur in any given case, and that in some instances even where rational treatment is early administered, a decided aggravation of the condition may follow unaccountably.

However, because of the economic element to be reckoned with, it is of some value to be able to give a fairly accurate prognosis in the handling of cases of lameness, as in the majority of instances the treatment and manner of after-care are determined largely by the expense that any prescribed line of attention will occasion.

A case of acute bone spavin in a horse of little value is not generally treated in a manner that will incur an expense equivalent to one-half the value of the subject. The fact is always to be considered in such cases, that even where ideal conditions favor proper treatment, the outcome is uncertain. Where less than six weeks of rest can be allowed the animal, one affected with bone spavin would therefore not be treated with the expectation of obtaining good results, as six weeks' time, at least, is necessary for a successful outcome. If the cost attending the enforced idleness of an animal of this kind is considered prohibitive for the employment of proper measures to affect a cure, and if lameness is slight, the animal should be given suitable work, but in cases of articular spavin in aged subjects, they should be humanely destroyed and not subjected to prolonged misery.

A thorough knowledge of the structure and functions of the affected parts is necessary to proceed in cases of lameness; likewise, the age, conformation and temperament of the subject need to be taken into consideration; the presence or absence of complications demand the attention; the kind of care the subject will probably receive directly influences the outcome; and the character of service expected of the subject, too, needs to be carefully considered before the ultimate outcome may reasonably be foretold.

The practitioner is often confronted with the problem of how best to handle certain cases. Will they do better under conditions where absolute quiet is enforced, or is it preferable to allow exercise at will? The temperament of the animal must be considered in such cases, and if a lame horse is too active and playful when given his freedom, exercise must be restricted or prevented, as the case may require. In cases of strains of tendons, during the acute stage, immobilization of the affected parts is in order. In certain sub-acute inflammatory processes or in instances of paralytic disturbance where convalescence is in progress, moderate exercise is highly beneficial.

Consequently, each case in itself presents an individual problem to be judged and handled in the manner experience has taught to be most effective, appropriate and practical, and the veterinarian should give due consideration to the comfort and welfare of the crippled animal as well as to the interests of the owner.

SECTION I.

ETIOLOGY AND OCCURRENCE.

In discussions of pathological conditions contributing to lameness in the horse, cause is generally classified under two heads— *predisposing* and *exciting*. It becomes necessary, however, to adopt a more general and comprehensive method of classification, herein, which will enable the reader to obtain a better conception of the subject and to more clearly associate the parts so grouped descriptively.

Though *predisposing* factors, such as faulty conformation, are often to be reckoned with, *exciting* causes predominate more frequently in any given number of cases. The noble tendency of the horse to serve its master under the stress of pain, even to the point of complete exhaustion and sudden death, should win for these willing servants a deeper consideration of their welfare. Too frequently are their manifestations of discomfort allowed to pass unheeded by careless, incompetent drivers lacking in a sense of compassion. Symptoms of malaise should never be ignored in any case; the humane and economic features should be realized by any owner of animals.

In the consideration of group causes, lameness may be said to originate from affections of bones, ligaments, thecae and bursae, muscles and tendons, nerves, lymph vessels and glands, and blood vessels, and may also result from an involvement of one or several of the aforementioned tissues, caused by rheumatism. Further, affections of the feet merit separate consideration, and, finally, a miscellaneous grouping of various dissimilar ailments, which for the most part, do not directly involve the locomotory apparatus but do, by their nature, impede normal movement.

AFFECTIONS OF BONES.

The bony column serving as the framework and support of the legs, probably constitutes the most vital element having to do with weight bearing and locomotion, and therefore during the acute and painful stage of bone affections, the pain becomes more intense in the process and pressure of standing than when the member is swung or advanced.

Certain bones are so well protected by muscular structures that they are not frequently injured except as a result of violence which may produce fracture. However, there are certain bones which receive the constant shock of concussion when the animal is subjected to daily, rapid work on hard road surfaces. Splints, ringbones and spavins are the most general examples produced by these conditions.

Varying pathological developments often result from concussion, contusion or other violent shocks to the bony structures. In such cases there either follows a simple periostitis which may resolve spontaneously with no obvious outward symptom, or osteitis, which may occur with tissue changes, as in exostosis; or the case may produce any degree of reaction between these two possible extremes.

Rarefying Osteitis, or Degenerative Changes.

Certain bone affections, such as osteomalacia or osteoporosis, are in the main, responsible for distortions and morphological changes of bone, causing lameness, permanent blemish and even resulting in death of the affected animal. The climatic conditions in some localities favor these occurrences but they may also be ascribed to improper food constituents and to possible infective agencies.

Rarefying degenerative changes manifested by exostosis involving the phalanges of the young, causing ringbone, are fairly common in occurrence throughout this country. This is due, supposedly, to a lack of mineral substance in the bony structure of the affected animals, and is known as rachitis—commonly called rickets. Since the affected subjects suffer involvement of several of the extremities at the same time, the theory of rachitic origin seems well supported.

Fractures.

Fractures of bones constitute serious conditions and are always manifested by lameness. A sub-classification is essential here for the student of veterinary medicine who would comprehend the technic of reduction and subsequent treatment in such cases.

Fractures are classified by many authorities as being *simple, compound*, and *comminuted*. This method is practical because it separates dissimilar conditions. There are also grouped fractures, the pathologic anatomy of which is similar. Classification on an etiological basis would attempt to associate conditions, the morbid anatomy and gravity of which would justly preclude their being combined.

Simple Fracture is a condition where the continuity of the bone has been broken without serious destruction of the soft structures adjacent, and where no opening has been made to the surface of the flesh. Such fractures do not reduce the bone to fragments. Long bones are frequently subjected to simple fracture, while short thick bones, such as the second phalanx, may suffer multiple or comminuted fractures.

Compound Fracture designates a break of bone with the destruction of the soft tissues covering it, making an open wound to the surface of the skin. This form of fracture is serious because of the attendant danger of infection, and in treatment, necessitates special precaution being taken in the application of splints that the wound may be cared for without infection of the tissues. These fractures generally occur as a result of some forceful impact through the flesh to the bone, or where the bones are driven outward by the blow. Common examples are in fractures of the metacarpus and metatarsus of the first phalanx. This kind of injury in mature horses usually produces an irreparable condition, and viewed economically, is generally considered fatal.

Comminuted Fractures, as the term implies, are those cases wherein the bone is reduced to a number of small pieces. This kind of break may be classified as simple-comminuted fracture when the skin is unbroken, and when the bone is exposed as a result of the injury, it is known as a compound-comminuted fracture. Such fractures are caused by violent contusion or where the member is caught between two objects and crushed.

Multiple Fractures.

Fractures are called *multiple* when the bone is reduced to a number of pieces of large size. This condition differs from a comminuted fracture in that the multiple fracture may break the bone into several pieces without the pieces being ground or crushed, and the affected bone may still retain its normal shape.

Further classification is of value in describing fractures of bone with respect to the manner in which the bone is broken—the direction of the fissure or fissures in relation to its long axis.

A fracture is *transverse* when the bone is broken at a right angle from its long axis. Such breaks when simple, are the least trouble to care for because there is little likelihood that the broken ends of bone will become so displaced that they will not remain in apposition. *Simple transverse* fracture of the metacarpus, for instance, constitutes a favorable case for treatment if other conditions are favorable.

Oblique fractures, as may be surmised, are solutions of continuity of bone in such manner that the fissure crosses the long axis of a bone at an acute or obtuse angle. These fractures are prone to injure the soft structures adjacent, and are frequently compound, as well. Moreover, because of the fact that the apposing pieces of bone are beveled, the broken ends of bone are likely to pass one another in such a way as to shorten the distance between the extremities of the injured member. Contraction of muscles also tends to exert traction upon a bone so fractured, resulting in a lateral approximation of the diaphysis and thus preventing union because the broken surfaces are not in proper contact.

Fractures are *longitudinal* when the fissure is parallel with the long axis of the bone. This variety of break is not infrequent in the first phalanx; and a vertical fracture of the second phalanx is also said to be longitudinal, however, there is little difference (if any, in some subjects) between the vertical and transverse diameters of this particular bone.

Green stick fractures are essentially those resulting from falls to young animals. They are usually sub-periosteal and when the periosteum is left intact or nearly so, no crepitation is discernible. If this fracture is *simple*, prompt recovery may be expected. Bones of young animals, because they do not contain proportionately as much mineral substance as do bones of adults, are more resilient and less apt to become completely fractured. They are, however, subject to what is known as green stick fracture.

Impacted fractures are usually occasioned by falls. When the weight of the body is suddenly caught by a member in such manner as to forcefully drive the epiphyseal portions of bone into and against the diaphysis, *multiple longitudinal* fractures occur at the point of least resistance. Parts so affected undergo a fibrillary separation, increasing the transverse diameter of the bone; or if the impact has been sufficiently violent, the portion becomes an amorphous mass.

In a treatise on the subject of lameness, the bones chiefly concerned and most often affected must be especially considered. The shape and size of a bone when injured, determines in a measure, the course and probable outcome in most cases, but of first and greater importance is the function of the bone. A fracture of the fibula in the horse need not incapacitate the subject, but a tibial fracture is serious and generally proves cause for fatal termination. The body of the scapula may be completely fractured and recovery will probably result in most cases without much attention being given to the subject, yet a fracture of the neck of this same bone constitutes an injury of serious consequence. The difference in the function of different parts of this same bone, as well as its shape and mode of attachment, determine the gravity of the case; so it is in fractures of other bones with respect to the course and prognosis of the case — function is the important factor to be considered.

Next in importance is the age of the animal suffering fracture of the bone. Capacity for regeneration is naturally greater in a vigorous, young animal than in aged or even middle-aged subjects. A healthy condition of the bone and the body favor the process of repair in case of fracture, and prognosis may be favorable or unfavorable, depending upon these factors mentioned for consideration. Individuals of the same species, differing in temperament, may comport themselves in a manner that is conductive to prompt recovery, or to early destruction. This feature cannot be overestimated in importance, as it is sometimes a decisive element, regardless of other conditions. A horse suffering from an otherwise remediable pelvic fracture may be so worried and tortured by being confined in a sling that the case calls for special attention and care because of the animal's temperament. Sometimes, the constant presence of a kind attendant will so reassure the subject that it will become resigned to unnatural confinement, in a day or two. This precaution may, in itself, determine the outcome, and the wise veterinarian will not overlook this feature or fail to deviate from the usual rote in the handling of average cases. Recovery may be brought about in irritable subjects by this concession to the individual idiosyncrasies of such animals.

AFFECTIONS OF LIGAMENTS.

Ligaments which have to do with the locomotory apparatus are, for the most part, inelastic structures which are composed of white fibrous tissue and serve to join together the articular ends of bones; to bind down tendons; and to act as sheaths or grooves through which tendons pass, and as capsular membranes for retention of synovia in contact with articular surfaces of bones.

Ligaments are injured less frequently than are bones. Because of their flexibility they escape fracture in the manner that bones suffer. They are, however, completely severed by being cut or ruptured, though fibrillary fracture the result of constant or intermittent tensile strain is of more frequent occurrence.

Simple inflammation of ligaments is of occasional occurrence but, unless considerable injury is done this tissue, no perceptible manifestation of injury results. No doubt many cases wherein fibrillary fracture of ligaments (sprain) takes place some lameness is caused, but because of the dense, comparatively nonvascular nature of these structures, little if any manifestation, except lameness, is evident. And such cases, if recognized are usually diagnosed by excluding the existence of other possible causes and conditions which might also cause lameness.

Certain ligaments are subjected to strain more than are others and therefore, when so involved, frequently cause lameness. Examples of this kind are affections of the collateral (lateral) ligaments of the phalanges. Because of the leverage afforded by the transverse diameter of the foot, when an animal is made to travel over uneven road surfaces, considerable strain is brought to bear on the collateral ligaments of the phalanges. A sequel to this form of injury is a circumscribed periostitis at the site of attachment of the ligaments and frequently the formation of an exostosis—ringbone—results.

Where sudden and violent strain is placed upon a ligament and rupture occurs, the division is usually effected by the ligament being torn from its attachment to the bone. In such cases, a portion of periosteum and bone is usually detached and the condition may then properly be called one of fracture. In some cases of this kind recovery is tardy, because of the difficulty in maintaining perfect apposition of the divided structures, and reactionary inflammation is not of sufficient extent to enhance prompt repair. In fact, some cases of this kind seem to progress more favorably, when no attempt at immobilization of the affected member is attempted.

If some freedom of movement is allowed, acute inflammation resulting in nature's provisional swelling soon develops and repair is hastened because of increased vascularity. But where luxation of phalanges accompanies sprain, reposition and immobilization are necessary—that is if cases are thought likely to benefit by any treatment.

Luxations—Dislocations.

Luxation or dislocation is a condition where the normal relation between articular ends of bones has been deranged to the extent that partial or complete loss of function results. When a bone is luxated (out of joint), there has occurred a partial or complete rupture of certain ligaments or tendons; or a bone may be luxated when an abnormal or unusual elasticity of inhibitory ligaments or tendons obtains.

Luxations may be practically classified as *temporary* and *fixed*. In temporary luxations, disarticulation is but momentary and spontaneous reposition always results; while a fixed luxation does not reduce spontaneously but remains luxated until reposition is effected by proper manipulation and treatment. Fixed luxation may be of such character as to be practically irreducible because of extensive damage done to ligaments or cartilage. Where a complete luxation of the metacarpophalangeal joint exists, it is probable that in most cases sufficient injury to collateral and capsular ligaments has been done to render complete recovery improbable, if not impossible.

Temporary luxation of the patella is a common affection of the horse and fixed luxation of this bone also occurs. As a matter of fact, in the horse, patellar luxation is the one frequent affection of this kind.

As a rule, complete disarticulation immobilizes the affected joint and in most instances there is noticeable an abnormal prominence in the immediate vicinity—in patellar luxation, the whole bone. In other instances the articular portion only, of the affected bone is malpositioned. Usually, luxation and fracture may be differentiated in that there is no crepitation in luxation and more or less crepitation exists in fracture.

It is evident, when one considers the symptomatology and nature of the affection, that fixed luxation is usually caused by undue strain or violent and abnormal movement of a part. Joints having the greater freedom of movement are apt to suffer luxation more frequently.

Arthritis.

The study of arthritis in the horse is limited to a consideration of joint inflammations which, for the most part, are of traumatic origin. Unlike the human, the horse is not subject to many forms of specific arthritis — tubercular, gonorrheal, syphilitic, etc.

A practical manner of classification of arthritis is *traumatic* and *metastatic.*

Traumatic arthritis may result from all sorts of accidents wherein joints are contused. Such cases may be considered as being caused by direct injuries. Instances of this kind, depending on the degree of insult, manifest evidence of injury which ranges from a simple synovitis to the most active inflammatory involvement of the entire structure and adjacent tissues.

The reactionary inflammation which attends a case of tarsitis caused by a horse being kicked is a good example of the result of direct injury. Such cases, if the contusion is of sufficient violence, result in arthritis and periarthritis. In inactive farm horses, during cold weather, this condition becomes chronic, swelling remains for weeks after all lameness and pain have subsided and occasionally hyperthrophy is permanent.

Arthritis occasioned by indirect injury, such as characterizes joint inflammation from continuous concussion, is seen in horses that are worked at a rapid pace on city streets or other hard road surfaces. Such affections may be acute, as in some cases of spavin, but are usually inflammatory conditions that do not occasion serious disturbance when these affections become chronic. If the involvement persists with sufficient active inflammation, there may follow erosion of cartilage and incurable lameness. If extensive necrosis of cartilage takes place, the attendant pain will be sufficient to cause the animal to favor the diseased part and such immobilization enhances early ankylosis — nature's substitute for resolution in this disease.

Wounds invading the tissues adjacent to joints, when these wounds are of considerable extent, cause inflammation of such articulations by contiguous extension of inflammation. As long as an injury remains practically aseptic, or if infected and the septic process does not involve the joint proper by direct extension, no more serious disturbance than a simple synovitis will result. If, instead, a periarthritic inflammation is serious or destructive in character, the type of arthritis will be grave—even though due to an indirect cause.

Where a vulnerant body penetrates all structures and invades the interior of the joint capsule the result is that a more or less active disturbance is incited. The introduction of a sterile instrument into a joint cavity, under strict asepsis, where a perfect technic is executed, does not cause perceptible manifestation of the injury, if the opening so made is small—such as a suitable exploratory trocar makes. But a puncture made in a similar manner and with the same instrument without due regard to asepsis is likely to cause an infectious synovitis and arthritis usually follows.

A larger opening than is produced by means of an exploratory trochar may be made into a joint cavity, causing escape of synovia as it is secreted for days and even for weeks and no serious or permanent trouble is experienced in some cases. If the synovitis or arthritis remains non-infected and the wound, traumatic or surgical, is not too large, healing by granulation occurs, and the discharge of synovia ceases. However, if synovial discharge persists too long because of tardy closure of an open joint, there is great danger of infection gaining entrance into the synovial cavity, or in some instances, desiccation of endothelial cells of the articulation occurs, in areas, and the reactionary inflammation eventually results in ankylosis.

A small puncture which introduces into the synovial cavity infectious material of active virulence will cause an arthritis that is more serious, much more painful and more difficult to handle than is occasioned by a wound of moderate size, that affords ready escape of synovia even through the virulence of the infection be the same.

Synovia is a good culture medium and the environment is ideal for multiplication of bacteria; consequently, the grave disturbances which may attend the introduction of pathogenic organisms into a synovial cavity as the result of a puncture wound are not to be forgotten. The veterinarian is in no position to estimate the virulency of organisms so introduced; neither can he determine the exact degree of resistance possessed by the subject in any given case. Therefore, he is uncertain as to the best method of handling such cases where an injury has been recently inflicted and positive evidence of the existence of an infectious synovitis is not present. If one could determine in advance the degree of infection and injury that is to follow small penetrant wounds of joint capsules, it would then be possible to select certain cases and immediately drain away all synovia and fill the cavity by injection with suitable antiseptic solutions.

This offers a broad field for experimentation which will in time be productive of a radical change in the manner of treating such cases.

Metastatic arthritis is seen more frequently in colts or young animals than in mature horses and we here take the liberty of classifying with the arthritis of omphalophlebitis and strangles the so-called rheumatic variety.

A specific polyarthritis or synovitis which attends navel infection of foals is perhaps the most frequent form of arthritis that is to be considered metastatic. This condition is truly a disease of young animals and, while it is a specific arthritis, the cause is yet to be attributed to any definite pathogenic organism with certainty. This condition is well defined by Bollinger as quoted by Hoare,[1] when he calls it a purulent omphalophlebitis due to local infection of the umbilicus and umbilical vessels, by pyogenic organisms, causing a metastatic pyemia.

This affection is grave; its course is comparatively brief; the prognosis is usually unfavorable; and omphalophlebitis occasions a form of lameness which at once impresses the practitioner that serious constitutional disturbance exists. Its consideration properly belongs to discussions on practice or obstetrics and diseases of the new born, and it has received careful attention and is discussed at length in these works.

A second form of metastatic arthritis is met with in strangles. Strangles occurs in the young principally and is not a frequent cause of synovitis or arthritis in the adult animal.

Strangles or distemper is, according to most pathologists, due to the Streptococcus equi. Hoare[2] states that in this type of specific arthritis the contagium is probably carried by the blood. He gives it as his opinion that even laminitis has occurred as a result of the streptococcus-equi. This, indeed, would point toward probable extension by the blood as well as by way of lymph vessels.

Septic synovitis and infectious arthritis are always serious affections even in young animals and much depends upon individual resistance and early rational treatment in such cases, if recovery is to follow.

The same general plan of treatment is indicated in this kind of septic synovitis as is employed in all cases of infective synovitis and septic infection in open joints. There is to be considered, however, the fact that the young animal is more agile, a better self-nurse, and in a general way more apt to recover than is the adult, under similar conditions.

Rheumatic arthritis, if one is justified in classifying rheumatic inflammation of joints as a metastatic form of arthritis, is not a common condition, though seen in mature and aged animals. Cases that may be diagnosed with certainty are usually advanced affections wherein dependable history is obtainable and the symptoms are well marked.

Rheumatism may be thought of, with respect to arthritic inflammation caused thereby, as a sort of pyemia. Undoubtedly, exposure to wet and cold weather is an active factor, but probably a predisposing one only. Likewise a member that suffers from chronic inflammation due to recurrent injury or to constant or repeated strain is less able to resist the vicissitudes of climate and work.

Consequently, rheumatic arthritis is to be seen affecting horses that are in service, more often at heavy draft work where they are exposed to severe straining of joints; where stabling is insanitary; and where they are obliged to lie down (if they do not remain standing) upon cold and wet ground or upon hard unbedded floors or paving.

Where such inhumane and cruel treatment is given animals those responsible ought to be impressed with the unfairness to the animal as well as the economic loss occasioned by inflicting such unnecessary and merciless treatment upon their helpless and uncomplaining subjects. The very nature of the veterinarian's work affords him constant and frequent opportunity to convince those who are responsible for keeping animals in this manner, that it is inhumane and unprofitable.

Cases of this kind are not uncommon about some grading and lumbering camps and in contract work where, often, shelter for animals is given little thought; the result is a cruel waste of horse-flesh.

Chronic articular rheumatism is occasionally observed in young animals that have never been in service. In these cases it seems that there exists an individual susceptibility and in some instances the condition is recurrent. Each attack is of longer duration, and eventually death results from continued suffering, emaciation and intoxication.

AFFECTIONS OF BURSAE AND THECAE.

Acute bursitis and thecitis is of frequent occurrence in horses because of direct injury from contusion, punctures and other forms of traumatism. These synovial membranes, with few exceptions, when inflamed occasion a synovitis that may be very acute, yet there is less manifestation of pain than in arthritis.

It is only in structures such as the bursa intertubercularis or in the sheath of the deep digital flexor that an inflammation causes much pain and is apt to result in permanent lameness. This is due to the peculiar character of the function of such structures.

An acute inflammation of a small bursa may even result in the destruction of such synovial apparatus without serious inconvenience to the subject, either at the time of destruction or thereafter. Obliteration of the superficial bursa over the summit of the os calcis is not likely to cause serious inconvenience or distress to the subject unless it be due to an infected wound. Even then, with reasonably good care given the animal, recovery is almost certain. Complete return of function of the member and cessation of lameness takes place within a few weeks in the average case.

Where an infectious synovitis involves a structure such as the sheath of the tendon of the deep digital flexor (perforans) the condition is grave and because of the location of this theca the prognosis is not much more favorable than in an articular synovitis.

Inflammation of bursae and thecae may be classified on a chronological basis with propriety because the duration of such affections, in many cases, materially modifies the result. A chronic inflammatory involvement of a theca through which an important tendon plays may cause adhesions to form. Or there may occur erosions of the parts with eventual hypertrophy and loss of function, partial or complete.

However, in general practice a classification on an etiological basis is probably more practical and we shall consider inflammation of bursae and thecae as *infectious* and *noninfectious*.

Infectious bursitis and thecitis is usually the result of direct introduction of septic material into the synovial structure by means of
injuries. Infection by contiguous extension occurs and also metastatic involvement is met with occasionally.

The noninfectious inflammation of bursae and thecae usually result from contusions or strains and generally run their course without becoming infective in character, where vitality and resistance of
the subject are normal.

In a general way, inflammation and other affections of bursae and
thecae are considered very similar to like affections of joints.

AFFECTIONS OF MUSCLES AND TENDONS.

Muscles and tendons having to do with locomotion are more frequently injured than are any of the other structures whose function is to propel the body or sustain weight. This is due in part to the exposed position of muscles and tendons. They serve as a protection to the underlying structures and in this manner receive many blows the force and violence of which are spent before injury extends beyond these tissues.

Muscles of the breast, shoulder and rump are most frequently the recipient of injuries of various kinds. The abductors of the thigh are subjected to bruising when horses are thrown astride of wagon poles or similar objects. Thus in one way or another muscle injuries are occasioned and cause lameness.

Traumatic affection of muscles of locomotion may be surface or subsurface—subsurface with little injury done the skin and fascia, but with subsurface extravasation of blood and masceration of tissue. Puncture wounds wherein the vulnerant body is of small diameter, are observed, and they occasion deep seated infectious inflammation of the parts affected, with surface wounds that are often unnoticeable. Such injuries—puncture wounds—are always serious, and because of the fact that, there exists little evidence of injury at the time of their infliction, treatment is usually deferred several days and often infection has become quite extensive when the practitioner is consulted.

Where infective wounds of muscles of locomotion occur, the course and gravity of the affection are directly influenced by the proximity of the injury to lymph plexuses. For instance, injuries causing an infectious inflammatory involvement of the adductors of the thigh may result in a generalization of the infection by way of the inguinal lymph glands.

Large open wounds that extend deep into muscles, render inactive such structures, and even where division is not complete, the pain occasioned causes the subject to favor the part in every way possible. Contraction of muscular fibers of such parts increases pain and because of this fact groups of muscles are at times disabled because of injury done to one muscle. Instances of this kind are frequently seen where shoulder injuries, which affect but one muscle, exist; yet because of such injury a marked swinging-leg lameness is present.

Tendons, because of their inelasticity, are subjected to injuries peculiar to themselves. In addition to being affected as are muscles, wounds of many kinds are found to affect tendons—contusions, interference wounds, penetrant wounds, incised wounds and lacerations.

However, the commoner form of injury done tendons, is strain or sprain. Because of the sudden tensile strain brought to bear upon tendons in the shocks of concussion, as well as in propulsion of the body, there frequently occurs a rupture of fibers and this we know as sprain.

Sprains may be considered as fibrillary fractures of soft structures and since this form of injury is subsurface, and limited to fractional portions of tendons, the inflammation occasioned usually remains an aseptic one. Reaction to this form of injury is characterized by inflammation, the course of which is erratic and variable. In chronic inflammation of tendons, where animals are continued in service, the usual sequel is contraction, or shortening of these structures.

The degree of contraction as well as its import varies in different subjects and in the various tendons which may be affected. Contraction is a slow-going process that is progressive, gradually causing a decrease in the length of the affected structure and eventually rendering the animal useless.

The practice of applying shoes with extended toe-calks for the purpose of "stretching" contracted deep digital flexor tendons (flexor pedis perforans) cannot be too strongly condemned. While the addition of an extension such as is ordinarily employed to the toe of a shoe of this kind, prevents for a time, frequent stumbling in such cases, the increased tensile strain which is thus occasioned hastens further contraction and subjects animals so shod to much unnecessary pain.

AFFECTIONS OF NERVES.

Because of their being protected by other structures, nerve trunks, which supply muscles of locomotion, are not subjected to frequent injuries such as contusions. However, they do become injured at times and the result is lameness, more or less severe.

Lameness originating from nerve affection, may involve central structures as, for example, the spinal cord, medulla oblongata or parts of the brain. In making an examination of some lame animals it is necessary to distinguish between cases of lameness that are of central origin and marked by incoördination of movement, and disturbances caused by other affections. Tetanus in its incipiency should not be confused with laminitis involving all four feet, or with certain forms of pleuritis, when careful examination is made, yet, in a way, to one not trained, the clinical symptoms are similar.

Disturbances of nerve function are caused in a variety of ways. It is not within the scope of this work to discuss central nervous disturbances caused by ingestion of mouldy provender, or disturbances of the brain or cord occasioned by infectious diseases, but mention of the existence of such conditions is appropriate.

By direct injury the result of blows, certain nerves are injured and muscles supplied by such nerves are rendered inactive. Depending upon the nature and extent of an injury thus inflicted, so the manner in which the affection is manifested varies. The suprascapular nerve is rather frequently injured causing partial or complete loss of function of the structures supplied by this nerve, and abduction of the scapulohumeral joint naturally results.

In some cases of dystocia the obturator nerve, (or nerves, if the involvement is bilateral), becomes injured by being caught between the maternal pelvis and some dense part of the fetus. This results in paralysis of the adductors of the thigh if sufficient injury is done.

It is said that nerves become over-stretched and held tense, in certain positions in which animals are obliged to remain while cast in confinement such as in some instances where unusual methods of restraint are employed. When the fore feet are drawn backward in such manner that great strain is put upon the radial nerve, it suffers more or less injury, and this is followed by partial or complete paralysis which may be temporary or permanent.

Degenerative changes affecting nerves, as in other tissues, occur and more or less locomotory impediment will follow — this depending upon the nerve or nerves affected and the nature of such involvement. Tumors may surround nerves and eventually the nerve so exposed becomes implicated in the destructive process. Before degenerative changes take place in the nerve substance, in such cases, pressure may completely paralyze a nerve when it is so situated. Melanotic tumors in the paraproctal tissue in some cases, because of the large size of the new-growths, cause paralysis of the sciatic nerve. The author has seen one case of brachial paralysis occasioned by an enormous development of fibrous tissue involving the structures about the ulna.

AFFECTIONS OF BLOOD VESSELS.

Lameness caused by disturbances of circulation may be due to structural affection of vessels, or functional disorders of the heart, and in some instances, a combination of these causes may be active.

Direct involvement of vessels is the commoner form of circulatory disturbance which occasions lameness, and the most frequent cause is of parasitic origin. Sclerostomiasis with attendant arteritis, thrombus formation and subsequent lodgement of emboli in the iliac, femoral, or other arteries, causes sufficient obstruction to prevent free circulation of blood, and the characteristic lameness of thrombosis results.

Indirect injury to vessels may occur because of contused wounds and subsequent inflammation of tissues supplied by such vessels. If the injury be of sufficient extent, considerable extravasation of blood will take place and the painfully swollen parts necessarily impair locomotion. In such instances lymph vessels participate in the disturbance, and the condition then becomes one wherein lymphangitis is the predominant disturbing element.

Angiomatous tumors are occasionally found affecting horses' legs—usually the result of some injury; and because of their size or position, they mechanically interfere with function. Furthermore, when such tumors are located on the inner or flexor side of joints, enough pain is occasioned that affected animals show evidence of distress, usually by intermittent lameness.

Horses do not suffer from distension of veins as does man, that is, there is rarely to be seen a case wherein much disturbance from this source exists.

AFFECTIONS OF LYMPH VESSELS AND GLANDS.

Inflamed lymph vessels and glands, the result of various causes, is a rather common source of lameness of horses. When one considers the proportion of tissue that is composed of lymph vessels and glands, it is then obvious that inflammation of these structures should cause a painful affection of members, when so affected, and that marked lameness and, in some instances, general constitutional disturbance such as anorexia, hyperthermia and general circulatory disorder are to follow.

Lymphangitis is most frequently occasioned by the introduction of septic material into the tissues; consequently, infectious lymphangitis is more frequently observed than the non-infectious type.

Specific infectious forms of lymphangitis are seen in glanders and in strangles; infectious types of this disturbance are found in many instances where, initially, a localized or circumscribed infection has occurred—the contagium having been introduced by way of an injury. An example of this kind is to be seen in a wound perforating the tibial fascia, where the injury is inflicted by means of a horse being kicked by another animal shod with sharp shoe-calks. Cases of this kind invariably result in a septic lymphangitis, and frequently lymphadenitis also occurs, for the inguinal lymph glands are so situated that their becoming contaminated is almost certain.

The trite phrase that "the tissues are bathed in lymph" should make clear the reason for the frequent occurrence of infectious lymphangitis and lymphadenitis. Foreign substances, bacteria and their products, inorganic material and in fact, anything that is introduced into the tissues, if soluble or miscible, will be taken up and conveyed by the afferent lymph vessels and disseminated throughout the system—hence the constitutional disturbances so frequently thus caused.

A non-infectious type of lymphangitis is frequently seen in the heavy draft breeds of horses and in such cases one or both hind legs are involved—it is very seldom that the thoracic limbs become so affected. Law[3] refers to this ailment as "Acute Lymphangitis of

Plethora in Horse." When one takes into consideration that these cases so frequently occur in heavy draft animals that are not worked regularly, that the pelvic limbs are the ones involved, and that the disorder often runs a short course (recovery often taking place within two or three days, with no treatment given other than a purge, circulatory stimulants and walking exercise) it is plausible to ascribe the condition to idiopathic factors.

Admitting the frequency of non-infectious lymphangitis, the practitioner must not confuse this type with similar lymphatic inflammation occasioned by nail punctures of the foot. It is very embarrassing indeed to make a diagnosis of lymphangitis — expecting that the disturbance will terminate favorably and uneventually — and later to discover a sub-solar abscess caused by a nail prick in the region of the heel.

Recurrent attacks of this disturbance cause hypertrophy of the lymph vessels and in some cases lymphangiectasis. In old subjects used for dissection or surgical purposes, it is very evident that in the ones which have suffered from chronic lymphangitis there exists an excessive amount of sub-facial connective tissue, making subcutaneous neurectomies quite difficult in some instances.

A sequel of chronic lymphangitis is a condition known as elephantiasis. In such cases there occurs a hyperplasia of the skin and subcutaneous tissues, resulting in some instances, in the affected member attaining an enormous size. Sporadic cases of this kind are to be seen occasionally, and are apparently caused by repeated attacks of lymphangitis. The affection is not benefited by treatment, and while a horse's leg may become so heavy and cumbersome as to mechanically impede its gait, as well as to fatigue the subject when made to do service even at a slow pace, elephantiasis causes no constitutional derangement. The hind legs, in elephantiasis, are affected and a unilateral involvement is more often seen than a bilateral one. The legs may be enlarged from the extremity to the body, but ordinarily the affection does not extend higher than the hock or the mid-tibial region.

A chronic, progressive, hyperplastic-degeneration exists in some cases and the subjects are in time rendered unserviceable because of the burden of getting about encumbered by the affected extremity.

In other animals hyperplasia progresses for a time—until the parts become greatly enlarged and conditions apparently attain an immutable state. Nevertheless animals so affected may continue in service for years without being distressed.

AFFECTIONS OF THE FEET.

Lameness is very often due to affections of the feet, and in all foot diseases probably the most constant cause is injury inflicted in some manner. Resultant from injury, there frequently develops complications and the one most often seen is infection.

Because of the fact that the feet are constantly exposed to germ-laden soil and filth, if not actually bathed in such infectious materials, it naturally follows that septic infection of some part of the feet must be of frequent occurrence.

Subsequent to being obliged to stand in mud and other damp or wet media, exposure to desiccating influences such as stabling upon dry floors, or at service on hot and dry road surfaces causes the insensitive parts of the feet to become dry, hard and brittle. This favors "checking" of the protecting structures and it frequently results in the formation of large fissures which expose the underlying sensitive parts of the feet and lameness is the inevitable outcome.

The function of the feet—bearing the weight of the animal at all times when the subject is not recumbent, and in addition to this, the increased strain put upon them at heavy draft work, together with the concussion and buffeting occasioned by locomotion, make the feet susceptible to frequent affections of various kinds.

Being almost completely encased by a somewhat inexpansible and insensitive wall and sole, renders the foot subject to pathologic changes peculiar to itself. The very nature of the structure of the foot together with the function of the sensitive lamina is sufficient cause for an affection unlike that seen involving other tissues—laminitis.

An exhaustive consideration of foot affections is a study in itself and one that comes within the realm of pathologic shoeing; nevertheless, a practical knowledge of diseases of the foot is indispensable in the diagnosis of lameness wherein the foot may be at fault.

The peculiar nature of foot affections renders them difficult of classification on any sort of basis that is helpful in the consideration of this subject. Injuries are the most constant cause of foot lameness, yet one must admit that there results complications because of infection in most instances; and that in some cases the injury is slight—just enough to permit the introduction of vulnerant organisms into the tissues. Therefore, one might well classify affections of the feet as infectious and non-infectious. There can be grouped in the class of infectious affections such conditions as nail pricks, calk wounds and canker. In the class of non-infectious affections one may consider conditions such as laminitis, strain and fractures.

FOOTNOTES:

[1] A System of Veterinary Medicine by E. Wallis Hoare, F.R.C.V.S., Vol. I, page 519.

[2] Ibid, page 807.

[3] Vol. I, page 534, Veterinary Medicine, by James Law, F.R.C.V.S.

SECTION II.

DIAGNOSTIC PRINCIPLES.

To observe attentively is to remember distinctly. — Poe.

Before treatment is administered in constitutional disturbances resulting in disease, *cause* is logically sought; so, in order to handle effectively any case of lameness, it is necessary first to discover the source of the trouble and contributing conditions affecting the structures. Hence, diagnostic ability is the prime requisite; and a thorough knowledge of pathologic anatomy or of surgical technic is of little value if this knowledge is not applied with the insight of the trained diagnostician.

The cruel and unnecessary methods employed by those untrained for diagnostics, cannot be too vigorously condemned. For instance, the application of an active and depilating vesicant upon a large area on the gluteal or crural region, in a case where the practitioner "guesses" the condition to be one of "hip lameness," constitutes an exposition of gross ignorance, and at once stamps the perpetrator as a crude bungler without scientific insight whose works are no credit to his profession. How much better it would be, if the practitioner does not see fit to call in a competent consultant, to prescribe a suitable agent to be given internally, and to recommend complete rest for the subject.

In establishing a diagnosis in such cases, the student or practitioner seldom has recourse to laboratory assistance, and his work is done by means of physical examination; therefore, a thorough knowledge and a clear conception of the physiology of locomotion are essential. Memorizing nosological facts without an understand-

ing of underlying principles is of no more practical benefit for qualification as a diagnostician in cases of lameness, than is the employment of similar methods in the study of theory and practice. A
knowledge of the dosage of drugs does not in itself qualify one as
being competent to administer such therapeutic agents to a proper
effect. How much is a practitioner benefited by the knowledge that
a high temperature is usually present in septic intoxication, if he is
not possessed of a scientific understanding of anatomy, physiology,
bacteriology and pathology, as well as the principles of clinical diagnosis?

In order to determine the reasons for certain symptoms manifested by the subject, an analysis of these symptoms is the proper
method of procedure, insofar as this is possible. If one may reason
that an animal assumes a certain position while at rest to allow relaxation of an inflamed tendon or ligament, such a fact enables the
diagnostician to recall that this is indicative of some specific ailment. In acute tendinitis, the subject while at rest, maintains the
affected member in volar flexion because this position permits relaxation of the inhibitory apparatus, including the inflamed tendon.
Likewise, the various abnormal positions assumed,—adduction,
abduction, undue flexion or pointing—have their own significance
and are taken into account by the trained diagnostician in the course
of an examination.

In the examination of lame subjects, where the cause is not obvious, a systematic method of diagnosis is pursued even by the most
expert practitioners. In all obscure cases of lameness a methodical
and thoroughly practical examination of the animal according to an
established procedure is necessary to determine the nature and
source of the affliction.

Anamnesis.

The first thing to be given consideration in diagnosis is the fact that related history of the case is not always dependable, because of lack of accurate observation or wilful deceit on the part of the owner or attendant. The successful veterinarian soon acquires the faculty of obtaining information in a manner best adapted to his client, — either by direct interrogation or by subtle means of suggestion, and in this way he draws out evaded facts essential to his diagnosis. In time he learns to make allowance for misstatements made to shield the owner or driver and to hide the facts of apparent neglect or abuse that the subject may have experienced. A suppurating carti-laginous quittor, complicated by the presence of a large amount of hyperplastic tissue, cannot be successfully represented to be an acute and recently developed affection, where a trained practitioner is left to judge the validity of the statement.

In complicated conditions, where there is evident a chronic dis-turbance which could not be conceived as sufficient cause for a marked manifestation of lameness, accurate history of the case may be of great aid in arriving at a diagnosis. An aged animal, having recently become very lame, showing a small exostosis on the first phalanx, and with the history given that the osseous deposit was of long standing, should at once lead the veterinarian to seek the source of trouble elsewhere.

Visual Examination.

As in all diagnostic work, a careful visual examination of the subject should be made before it is approached. The novice is given to hasty examination by palpation, not realizing how much may be revealed by a careful scrutiny of the subject. In this way he is led to erroneous conclusions which the skilled diagnostician has learned from experience to avoid. *Too much emphasis cannot be placed on the importance of making a thoughtful visual examination in every instance before the subject is approached.* In this examination, type, conformation and temperament are taken into account at once, for each of these qualities is in itself, a determining factor in predisposing a subject to certain ailments or inherent attributes, which may exert a favorable or unfavorable influence upon existing conditions and thus make recovery probable or otherwise.

Draft animals are less likely to be permanently incapacitated as a result of tendinitis, than are thoroughbreds. Likewise, one would not expect to find this affection present in heavy harness horses as frequently as in light harness animals.

Mal-formation of a part, or an asymmetrical development of the body as a whole, may render an animal susceptible to certain affections which cause lameness. A "tied in" hock predisposes the subject to curb, and an animal having powerful and well-developed hips and imperfectly formed hocks, will, if subjected to heavy work, be a favorable subject for bone spavin.

The matter of temperament cannot be disregarded in diagnosis, for in some instances, it is the chief determining factor which materially influences the outcome of the case. A nervous, excitable animal, that is kept at hard work, may, under some conditions, be expected to experience disturbances which more lethargic subjects escape. Nervous subjects, it is known, are more prone to azoturia than are those of lymphatic temperament. Furthermore, the lymphatic subject often recovers from certain bone fractures which are successfully treated only when the animal is sufficiently resigned by nature to remain confined in a sling for weeks without resistance.

The physiognomy of a subject is often indicative of the gravity of its condition. The facial expression of an animal suffering the throes of tetanus, azoturia, or acute synovitis, is readily recognized by the experienced eye, and upon physiognomy alone, in many instances, may the opinions regarding prognosis be based. Particularly is this true where death is a matter of minutes, or at most is only a few hours distant.

Due allowance should be made for restiveness manifested by some more nervous animals when the surroundings are strange and unusual. In such instances, even pathognomic symptoms may be masked to the extent that little, if any, sign of pain or malaise is evinced. In these cases the subject should be given sufficient time to adjust itself to the new environment, or it should be removed to a more suitable place for examination. Animals quickly detect the note of friendly reassurance in the human voice and can very often be calmed by being spoken to.

By visual examination one may detect the presence of various swellings or enlargements, such as characterize bruises and strains of tendons where inflammation is acute. Inflammation of the plantar (calcaneocuboid) ligament in curb is readily detected when the affected member is viewed in profile. Spavin, ringbone, splints, quittor and many other anomalous conditions may all be observed from certain proper angles.

The fact that the skins of most animals are pigmented and covered with hair, precludes the easy detection of erythema by visual examination, consequently this indicator of possible inflammation is not often made use of in the examination of equine subjects.

Attitude of the Subject.

The position assumed while the subject is in repose, is often characteristic of certain affections and this, of course, is noted at once. The manner in which the weight is borne by the animal at rest, should attract the attention of the diagnostician and if the attitude of the subject is abnormal or peculiar, the examiner tries to determine the reason for it. If weight-bearing causes symptoms of pain, the affected member will invariably be favored and held in some one of a number of positions. The foot may contact the ground squarely and yet the leg may remain relaxed and free from pressure; volar flexion, in such cases, is indicative of inflammation of a part of the flexor apparatus. If the condition be very painful, position of the afflicted member is frequently shifted, but in all cases where the pain is not so keenly felt, the inflamed member is held in a state of relaxation. There is need then, for a knowledge of anatomy and certain principles in physics to enable the observer to determine just which structures are purposely eased in this manner. Where palpation of parts is possible, one does not need to depend on visual examination alone, and it is always wise to take into consideration every factor that may influence conditions. Manipulation or palpation of the structures thought to be involved, should not be resorted to until a careful and thorough observation of the subject has revealed all that it can reveal to the diagnostician.

In all conditions where extreme pain is manifested by the constant desire of the animal to keep its foot in motion off the ground, examination should be made for local cause. This is seen in certain septic inflammations of the feet such as those caused by nail punctures invading the navicular joint, or in newly made wounds where nerves have been divided and the proximal end of such a nerve is exposed to pressure or irritation.

"Pointing" affords a comfortable position in some cases of navicular disease, and in a unilateral affection, one may observe the subject bearing weight with one sound member, while the affected foot is planted well ahead of the sound one. In a bilateral involvement of this kind, weight may be frequently shifted from one foot to the other, or in chronic cases, where no marked pain is experienced, the subject stands squarely upon both front feet and no peculiar shifting of weight or pointing is evident.

In some cases of hip or shoulder involvement, complete relaxation of all parts of the affected member may be noticed. In brachial paralysis, the pectoral member is held limply; if the patient is made to move, it is evident there is lack of innervation to the afflicted part. In some cases where contusion has caused acute inflammation of the member, the subject instinctively tries to keep it inactive to relieve the pain which movement occasions.

Where there is an active and painful inflammation of the prescapular lymph glands and contiguous structures, in some cases of "levator-humeri abscess," the scapulohumeral joint is extended. This is brought about by flexion of the elbow and carpal joints.

There are some cases of bi-lateral affections which occasion such pain during weight-bearing that the subject shifts its weight from one affected leg to the other; an example of this condition may be observed in any acute case of gonitis which affects both patellar regions, making it equally painful to bear the weight on either member.

A peculiar characteristic position is assumed in acute laminitis of the fore feet. In such instances, the hind feet are brought forward under the body sufficiently to relieve the front feet of the weight, insofar as is possible by the abnormal position taken in cases of acute laminitis.

So in each position that is abnormal to any degree, assumed by a suffering animal, there may be deduced, the fact that the subject is attempting to relieve the affected structures, and in each clinical picture of this kind, the trained diagnostician sees some index to the nature and source of the trouble. Further examination is rendered more effective because of this preliminary visual examination which has precluded the unnecessary annoyance of the animal by manipulating unaffected structures.

It has been presupposed in the foregoing, that the one making visual examination of a lame animal for diagnostic purposes, will remember that with the normal animal the weight is borne equally well with both fore legs; and that this is done without shifting from one to the other; and that the pelvic limbs do not support the body in this manner. Normal subjects shift their weight from one hind leg to the other and the one relaxed, rests in a state of flexion with the toe on the ground and the heel raised.

Examination by Palpation.

In nearly every case where lameness exists an examination of the affected parts, by palpation or by digital manipulation, is necessary before an accurate conclusion may be drawn; but in making this kind of an examination one needs to exercise good judgment lest he fail to acquire a correct impression of the actual existent conditions. There is need for the diagnostician, here, as well as in other conditions where physical examination is made, to approach the subject in a manner that will not excite or disturb to the extent that the animal will, in one way or another, resist or object to the approach of the diagnostician, thereby masking the symptoms sought. The practitioner would best acquire skill as a horseman—if he is not possessed of such—and handle each individual subject in the manner calculated to best suit the temperament of the animal examined. The unbroken subject is not handled as satisfactorily as is the intelligent family horse; in the former, in some cases, little dependence is placed upon digital examination.

By palpation one is enabled to recognize hyperthermia and this, *in lieu* of dependable history, is at times sufficient evidence upon which to determine the duration of any given inflammatory affection.

By comparison of different parts of the same member or with an analogous portion of another member any marked increase in the apparently normal temperature of a part at once signalizes inflammation. In this manner, in examining a case where laminitis or other inflammation of the feet is suspected, one may arrive at a fairly accurate conclusion without the employment of other means. Throbbing vessels are not always easily recognized if the subject is a victim of chronic lymphangitis.

In some instances, where a moderate degree of lameness exists and cause is apparently obscure, the recognition of hyperthermia may be the deciding factor in establishing a diagnosis. In cases of sprained ligaments in the phalangeal region, because of the dense character of the structures involved, little if any evidence of the cause of lameness, other than local heat, may be found twenty-four hours after the injury has been inflicted.

In order to determine the amount or extent of hyperthermia with a fair degree of accuracy in any given case, one must make due allowance for external conditions affecting temperature; also the effect of a considerable amount of hair covering an area, as well as any possible dirt contacting the surface of the skin must be taken into account. All dirt should be removed if practicable, so that the diagnostician's palms may come as nearly in contact with the inflamed structures as possible. Then, too, the sense of touch if the operator's hands are chilled, is not dependable. In such instances the novice will need to be deliberate as to his findings—whether or not hyperthermia really exists. Such an examination is of little value where the subject's feet are wet and an examination is hurriedly made, as in cases of suspected laminitis.

Often, before being able to distinguish the presence of a hyperthermic condition, one is impressed with the fact that an animal manifests evidence of being supersensitive. In fact, some animals in the anticipation of pain at the touch of an injured part, will instinctively withdraw—in self-protection—such an ailing member or resist the approach of the practitioner. This sensitiveness is more apparent in animals that have been subjected to previous manipulation or treatment which has occasioned pain, and consequently, allowance must be made for this exhibition of fear. No better example of this condition can be imagined than is present in cases of "shoe boil," where there exists an extensive area of acute inflammation of the elbow. There is always more or less surface disturbance wherever vesication has been produced, and in cases where irritants of any kind have been employed for several days or a week previous to an examination, more or less supersensitiveness is to be expected.

One must not lose sight of the fact that unscrupulous dealers,—"traders"—make use of their knowledge of this principle in various way usually for the purpose of attracting attention to a part, which, presumably might have been blistered in order to intentionally produce inflammation of tissues, in this way, causing lameness which is not manifested until an animal has been kept by its new owner for twenty-four hours or more. This, to be sure, usually makes a dissatisfied purchaser who is willing to dispose of his newly acquired animal at a sacrifice, thus enabling the original owner or his agent to regain possession of the victimized animal at less than its real value.

Some nervous animals, because of the manner of approach of the practitioner, are wont to flinch, and there is manifested a pseudo-supersensitiveness. Young animals not accustomed to being handled are likely to be timorous, and one must not hastily conclude that a part is painful to the touch because the subject resents even gentle digital manipulation of such parts. In instances of this kind, one needs to compare sensibility by manipulation of different parts of the subject's body in a careful and gentle manner; and by exercising patience and good judgment in such work, it is possible to actually distinguish between normal sensibility and abnormal sensitiveness, in most cases. Here, again, the diagnostician needs to possess skill as a horseman and good judgment as to individual temperament of different animals, under any condition which may exist at the time he makes his examination.

By palpation alone, one can recognize the presence of fluctuating enlargements; one may not only recognize such conditions, but distinguish between a fluctuating mass such as exists in non-strangulated hernia and a large fibrous tumor. By palpation, for the recognition of density and for determining the presence or absence of hyperthermia, one may decide that there exists an abscess and not a tumor. Edematous swellings are recognized by palpation,— the characteristic indentations which may be made in dropsical swellings are pathognomonic indicators. In this manner it is easy to differentiate post-operative or post-traumatic edemas which may or may not cause lameness. At any rate, it is essential to take into account all determinate conditions that may assist in the prognosis of any given case, for the purpose of being able to outline rational remedial measures. To be able to distinguish between the generalization of a septic infection in its incipiency, and a more or less benign edema, is largely possible by digital manipulation alone. An extremity may be greatly swollen because of the existence of chronic lymphangitis, influenza, or an acute septic infection occasioned by the introduction of pathogenic and aerogenic organisms. Since the effect produced by these dissimilar ailments are productive of conditions that may terminate favorably or unfavorably, it becomes necessary for the diagnostician to develop a trained, discriminating, tactile-digital sense, in order to correctly interpret existing conditions, and handle cases in a rational and skillful manner.

In order to ascertain the extent and exact location of a tumor, an exostosis, or other enlargements, the diagnostician, here also, needs to be in possession of a trained tactile sense and in addition if he be fortified with an accurate knowledge of normal anatomy and pathology, he is able to arrive at proper conclusions, when digital manipulations have been employed. Fibrous tumors are sometimes located in the inferior part of the medial side of the tarsus—exactly over the seat of bone-spavin. Such tumors, when the affected member is supporting weight, are not to be distinguished from exostoses; but as soon as the affected leg ceases to bear weight, it may be passively flexed and the nature of the enlargement recognized because it may be slightly displaced by digital manipulation. Displacement, of course, is not possible with an exostosis.

A necessary qualification, which the diagnostician must possess, is that of being able to judge carefully the nearness of any given exostosis to articular structures. Also, the extent or area of the base of an exostosis as well as its exact position, needs be determined before one may estimate the probable outcome in any case,—whether treatment should be encouraged or discouraged by the practitioner. Periarticular ringbone may, because of the size and location of the exostosis, constitute a condition which cannot be relieved in any way in one case, and in another, because of the manner of distribution of such osseous deposits, the condition may be such that prompt recovery will follow proper treatment. In the examination of an exostosis of the tarsus, it is particularly important to determine the exact location of the exostosis—whether or not the spavin involves the tibial tarsal (astragulus) bone very near its tibial articular portions. Obviously, if articular surfaces of joints are involved, complete recovery cannot result despite the most skillful attention given the subject.

Passive Movements.

Wherever it is possible to gain the confidence of a tractable animal to the extent that it will relax the structures sufficiently to make possible passive movement of affected parts, much is to be learned as a result of such manipulation. By this method one may differentiate true crepitation, false crepitation, luxation and inflammation of ligaments that have been injured, as in sprains of such structures in the phalangeal region.

True crepitation is recognizable by the characteristic vibration which is interpreted by tactile sense. It is possible to recognize fracture by the use of other methods—auscultation, tuning fork tests, etc., but in ordinary veterinary practice one must rely upon the sense of touch for recognition of crepitation.

Where pain is not so great that relaxation of parts does not occur, one can, by gently moving an extremity in various directions—as in flexion, extension and lateral motion as well as by rotation—cause to be manifested this peculiar grating,—the friction of newly broken bone. This is known as *true crepitation*. Where the subject, suffering phalangeal fracture, manifests evidence of pain due to tensing the structures about a fractured part, one may anesthetize the parts by using about two cubic centimeters of a two per cent. solution of cocain upon the plantar nerves, proximal to the fracture. It is perhaps best to deposit the cocain solution by means of two hypodermic punctures at different points along the course of each nerve, though closely situated to one another, thereby making more sure of the solution actually contacting the nerve. In some multiple fractures of the first or second phalanx this is quite necessary; otherwise, pain produced by passive manipulation causes the subject to keep the tendons so tense that crepitation may not be detected. The unnecessary infliction of pain is always to be avoided.

We know as *false crepitation* a vibrating impulse occasioned by normal contact of articular portions of bones such as in the metacarpophalangeal joint when this structure is passively moved, where the subject permits the parts to remain in a state of complete relaxation.

Attempts to recognize supersensitiveness or inflammation by means of passive movement of the shoulder or hip, whether gently or forcefully, is not productive of good, in any case, in large animals. Because of the bulk and weight of parts so manipulated, as well as the resistance the subject offers even in normal cases, no accurate conclusion is to be arrived at in this manner in the average instance. Animals nearly always resist the placing of members in any position that is so unusual and uncomfortable as that which is required to materially displace the component tissues of the shoulder or hip; therefore, such practice is useless because one can not distinguish between normal resistance and flinching caused by painful sensations in injured parts. Such manipulations are practical in small animals.

Observing the Character of the Gait.

In order to determine the degree of lameness as well as its character, it is necessary to cause the subject which is being examined, to move in some manner. The degree of inconvenience or distress experienced by a lame animal that is being so examined is manifested by the character of the claudication; and where much pain is occasioned in locomotion there is disturbance of respiration; perspiration may be noticeable and in some instances manifestation of nervous shock are very evident—this in timid, nervous animals that anticipate being punished when approached and, consequently, make every effort possible to move when urged to do so. An animal, then, should be moved only sufficiently to cause it to exhibit the degree of lameness present in any given case, and if a marked impediment is manifested it is not necessary to cause the subject to be exerted to the extent of inflicting, in such manner, unnecessary punishment. Further or conclusive examination is made by palpation. To cause the subject to move, an assistant may simply lead the animal with a halter and compel it to walk a few steps. In this way, lameness, whether manifested during the weight-bearing period of an affected member, or when such a member is being advanced, or whether a combination of the two conditions exists, is made apparent. In the words of Dollar, one is thus enabled to recognize the existence of "supporting-leg-lameness," "swinging-leg-lameness" or "mixed lameness."

When the cause of lameness is not strikingly apparent it becomes necessary to have the subject moved farther than a few steps and at different paces. Depending then, upon the character of lameness manifested, as well as upon its degree of intensity, one needs to exercise the subject in various ways, but this should not be overdone.

The first thing apparent in the lame subject in action, is the lame leg. If this is not readily determinable, as in some complicated cases, the leg or legs which are at fault are to be discovered by further examination, and to do this,—word-pictures convey little that is helpful in difficult cases,—long practice is the one route by which one may become efficient; that is, by experience gained after fundamental principles in the diagnosis of lameness have been mastered.

For a careful study of supporting-leg-lameness involving a fore limb, the subject is driven or led *toward* the one making such examination. If a hind leg is to be observed, the animal is made to travel *away from* the examiner. Where there exists swinging-leg-lameness, the subject should be caused to move past the diagnostician, so that he may get a side view of the subject while it is in motion.

In every case such examinations are made to the best advantage if the practitioner can view his patient from a little distance. Here, again, a visual examination is made but this cannot be successfully executed, in difficult cases, if the practitioner is stationed at too close range.

The average subject is best observed by being led, rather than being ridden, and in so doing the animal should be given moderately free rein. A close grasp on the lead may interfere somewhat with head movements. Nodding of the head with the catching up of weight by a sound member in supporting-leg-lameness of a fore leg, constitutes the chief symptom considered in detecting the lame leg.

Where supporting-leg-lameness affects a hind limb the head is raised at the time weight is caught by the sound member—here the long axis of the subject's body may be likened unto a lever of the first class. The posterior part of the body, at the time weight is taken upon the sound leg, is as the long arm: the fore limbs the fulcrum, and the subject's head the weight, which is lifted. The head movements of a horse at a trot, in supporting-leg-lameness of a front leg, synchronize with the discharge of weight from a lame leg to the opposite one if sound; but in pelvic limb affections, the head is thrown or jerked upward as weight is caught by the sound member,—this peculiar nodding movement is *opposite* in the two instances.

In pacing horses, since front and hind legs of the same side are advanced at the same time, there occurs in supporting-leg-lameness, a nodding of the head with discharge of weight from the lame leg, and a dropping of the hip as weight is caught by the sound pelvic member. In observing animals that are limping, (as in supporting-leg-lameness) one notices particularly the sacro-iliac region in hind leg affections and the occipital region in lameness of the front legs.

Where there exists a bilateral affection, (such as characterizes some cases of navicular disease or other affections causing supporting-leg-lameness) there occurs no nodding of the head; weight is supported for an equal length of time upon each one of the two legs, but the stride[4] is shortened. The gait, in such cases, is peculiar, animals appearing stiff and they are said, by horsemen, to have a "choppy" gait.

It is desirable, in some cases, to cause an animal to move from side to side; in other instances the subject is best made to walk or trot in a circle, and if the circle be very small the animal then particularly employs the inner fore leg as a pivotal supporting member. To augment the manifestation of certain affections, it is necessary to cause the patient to walk backward, and each one of these tests of locomotion serves to point out in a more or less characteristic manner, the site of the affection which is causing lameness in different cases.

Sprains or injuries of lateral ligaments of the extremities, ringbone and certain foot affections, are made manifest by a side to side movement or a pivotal movement. In fact, wherever it is possible to cause undue or unusual tension to be exerted upon an inflamed structure, manifestation of pain is the response. In an inflamed condition of the lateral side of the phalanges, unequal weight-bearing such as a rough road surface will, by virtue of the leverage which the solar surface of the foot affords, cause undue strain upon such inflamed parts, and increased lameness is evident.

When an animal is made to travel in a circle, when a member affected with supporting-leg-lameness is on the inner side of the circle, lameness is accentuated because weight is borne by the lame leg for a greater length of time, the result of such circuitous manner of locomotion. In swinging-leg-lameness, on the other hand, because pain is increased at the time an affected member is being advanced, lameness is increased when the subject is made to travel in a circle, with the lame leg on the outside of a circle thus described.

In supporting-leg-lameness, the transientness of the weight-bearing period upon the affected member is the determining factor in the production of lameness. This unequal period of weight-bearing upon the front legs, for instance, causes an acceleration in the advancement of the sound member, in order to relieve the diseased one which is bearing weight. In other words, when an animal that is affected with supporting-leg-lameness travels in a straight line, since weight is borne by the diseased leg for an abnormally short period of time, the sound member needs be in the act of advancement a correspondingly short period. The result is then, an unequal division of stride; a nodding of the head with the catching up of weight by the sound leg, — in front leg affections — and this is termed *limping*.

With continuous exertion as in travel for a considerable distance, in some cases, lameness becomes less evident — as in spavin. This "warming out" process is due in a measure to the parts becoming less sensitive upon exertion, and is to be seen, to a limited extent, in all inflammatory affections that are not too severe; consequently, in some cases, examination of a lame animal should begin in the stall, for in instances where the impediment is not marked, there may be no evidence of lameness after the subject has walked a few steps. In other cases, lameness increases as the subject continues to travel, and often to the extent that the impediment becomes too severe to allow the animal being serviceable. Therefore, one can not, in every case of lameness observed, positively determine the gravity of the situation, without having seen the affected animal in action for a sufficient length of time to understand the nature of the condition existing. This necessitates driving the animal for several miles in certain cases.

Sometimes it is impossible to arrive at any definite conclusion, as the result of a single examination, and it then becomes necessary to see the subject again at a later date, or under more favorable circumstances. This is to be expected in some conditions where there exists rheumatic affections, and also in some foot diseases.

In the examination of young animals, unused to harness and to other strange incumbrances, one is obliged to make allowance for impediments of gait, which are not occasioned by diseased conditions. Such affections have been termed "false lameness." Young mules that are not well broken to harness, are difficult subjects for examination and in some cases it is necessary to have them led or driven for a considerable distance before one can definitely interpret the nature of the impediment in the gait when lameness is not pronounced. It is especially difficult to satisfactorily examine such subjects, for the reason that their normal rebellious temperaments cause resistance whenever a strange person approaches them, as it is necessary to do for an examination by palpation. In such cases—if an examination does not reveal the cause of trouble, rest must be recommended and further examination made at a later date, whereupon any new developments may be noted, if such changes exist.

Special Methods of Examination.

After having completed a general examination of a lame animal—obtaining the history of the case, noting its temperament, type, size, conformation, position assumed while at repose, swellings or enlargements if present, causing the subject to move to note the degree and character of lameness manifested; palpating and manipulating the parts affected to acquire a fairly definite notion of the nature of an inflammation or to recognize crepitation it becomes necessary in some cases to employ peculiar means of examination in singular instances. This may be done by making use of cocain in solution for the production of local anesthesia as in lameness of the phalanges. Such means are not, in themselves, dependable but are valuable when used in conjunction with all other available and practical methods.

Trial use of various shoes in order to shift the weight from one part of the foot to another or to cause an animal to "break over" in a different manner so that the gait may be changed, constitutes a special test procedure. The use of hoof testers or of a hammer to note the degree or presence of supersensitiveness is another means that is of practical service. No examination, in any case of lameness, is complete without having removed the shoe and scrutinized the solar surface of the foot.

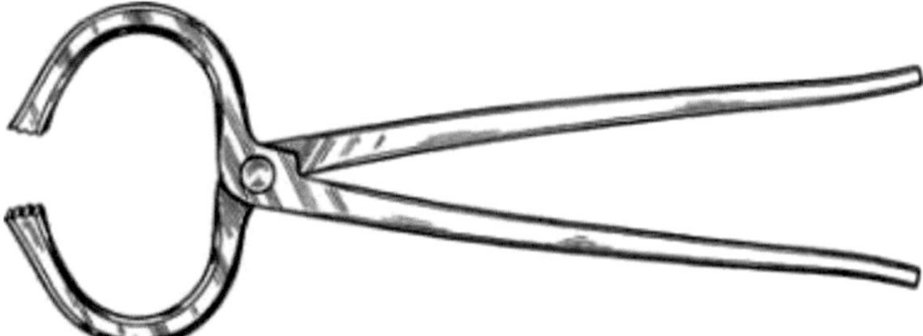

Fig. 1—Hoof testers with special jaws of sufficient size to grasp the largest foot.

Diagnosis by exclusion, finally, is resorted to, and, as in any other case where the recognition of cause is difficult, exclusion of the existence of conditions,—one at a time, by an analysis of symptoms—generally enables the practictioner to eliminate all but the disturbing element.

FOOTNOTES:

[4] By stride is meant the distance between two successive imprints of the same foot. The term is not used in this work as being synonymous with step.

SECTION III.

LAMENESS IN THE FORE LEG.

Anatomo-Physiological Review of parts of the Fore Leg.

For supporting weight, whether the subject is at rest or in motion, the bony column of the leg, together with attached ligaments, tendons and muscles, is wonderfully well adapted by nature for the function which they perform. The several bones which go to make up the supportive portion of the leg, are so joined at their points of articulation, that a minimum degree of strain is put upon each attachment.

The upper third of the scapula, with its cartilage of prolongation, is sufficiently broad and flattened that it fits snugly against the thorax without necessity for a complicated method of attachment—the clavicle being absent, attachment is muscular.

Smith[5] has very aptly stated that:

"It seems quite legitimate to regard the muscular union between the thorax and forelimb as a joint. There are no bones resting on each other, no synovia; but where the scapula has its largest range of movement there is a remarkable amount of areolar tissue, which renders movement easy. The whole central area beneath the scapula and humerus not occupied by muscular attachment, is filled with this easy-moving, apparently gaseously distended, crepitant, areolar tissue over which the fore legs glide on the chest wall as freely as if the parts were a large, well lubricated joint."

The scapulohumeral articulation (shoulder joint) is an enarthrodial (ball and socket) joint but because of its being held more or less firmly against the thoracic wall by muscular and tendinous attachment, and because a part of this attachment affords a means of support for the body itself, there is no need for binding ligaments and movement is possible in all directions even though restricted as to extent.

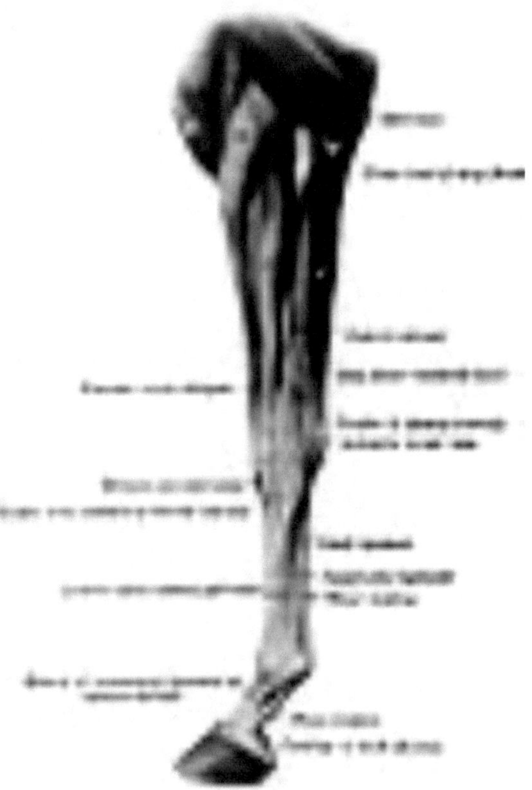

Fig. 2—Muscles of Left Thoracic Limb from Elbow Downward; Lateral (External) View.
a, Extensor carpi radialis; g, brachialis; g', anterior superficial pectoral; c, common digital extensor; e, ulnaris lateralis. (After Ellen-

berger-Baum, Anat. für Künstler.) (From Sisson's "Anatomy of the Domestic Animals").

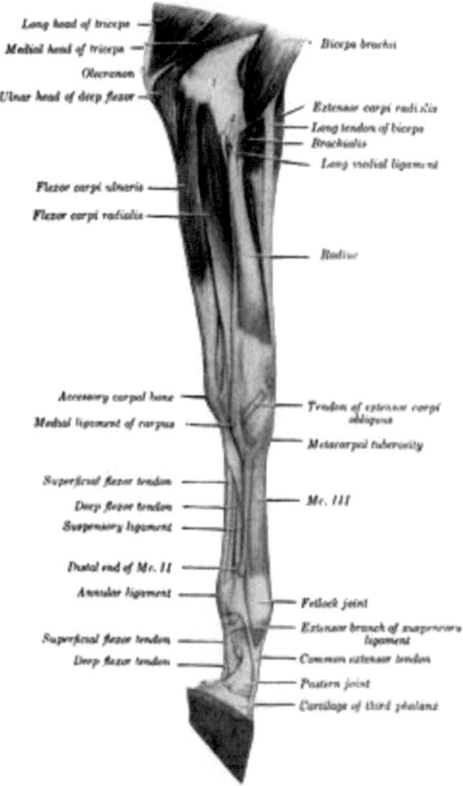

Fig. 3 — Muscles of Left Thoracic Limb from Elbow Downward; Medial (Internal) View.
The fascia and the ulnar head of the flexor carpi ulnaris have been removed. 1, Distal end of humerus; 2, median vessels and nerve. (From Sisson's "Anatomy of the Domestic Animals").

Undue extension, (by extension is meant such movement as will cause the long axis of two articulating bones to assume a position which approaches or forms a straight line — opposite to flexion), of the scapulohumeral joint is impossible while weight is borne, because of the normally flexed position of the humerus on the scapula;

whereas flexion, beyond desirable limits, is inhibited by the biceps brachii (flexor brachii or coracoradialis) muscle.

The distal end of the humerus, however, articulating with the radius and ulna in a fashion that no support is lent by any sort of contact with the body, is a ginglymus (hinge) joint and lateral motion, because of the long transverse diameter of its articular portions, is easily prevented by the medial and lateral ligaments (internal and external ligaments). Flexion of this, the humeroradioulnar joint (elbow), is restrained by the triceps brachii and extension is checked by the biceps brachii (flexor brachii).

The carpal joint (erroneously called the knee joint), is composed of the several carpal bones which interarticulate and, when taken as a group, serve as a means of attachment and articulation for the radius and metacarpal bones.

The transverse diameter of this joint is long, thus giving it contacting surfaces that are sufficiently extensive to minimize the strain upon the mesial and lateral ligaments (internal and external lateral common ligaments). Motion is that of flexion and extension; slight rotation is possible when the position is that of flexion. While supporting weight the carpus is fixed in position by a slight dorsal flexion, but undue dorsal flexion is prevented by the flexor muscles and tendons and volar-carpal or annular ligament, together with the superior check ligament.

The metacarpophalangeal articulation (fetlock joint), is a hinge joint and its articular surfaces contact one another, with respect to their having a long bearing surface from side to side, as do all ginglymus (hinge) joints. Two common lateral ligaments bind the bones together. While bearing weight, there is assumed a position of slight dorsal flexion, undue flexion being checked by the inhibitory apparatus of the joint—check ligaments, and their tendons and the suspensory ligament. The inhibitory apparatus of the fetlock joint is materially reinforced by the proximal sesamoid bones. Situated as they are, between the bifurcating portions of the suspensory ligament and the posterior part of the distal end of the metacarpus—with which they articulate—the sesamoid bones serve to change the course of the branches of the suspensory ligament in a manner that they give firm support to this joint. Volar flexion is limited by the extensors of the phalanges.

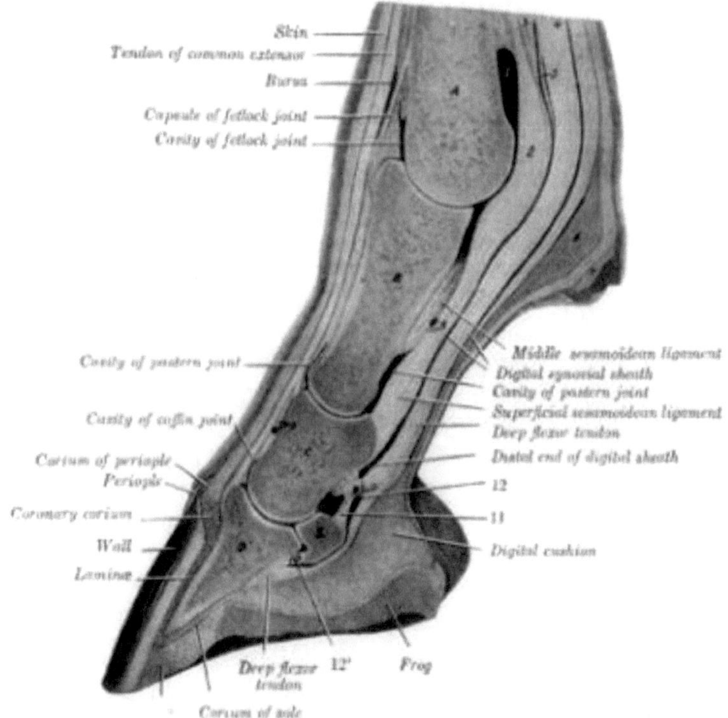

Fig. 4—Sagital Section of Digit and Distal Part of Metacarpus.
A, Metacarpal bone; B, first phalanx; C, second phalanx, D, third phalanx; E, distal sesamoid bone; 1, volar pouch of capsule of fetlock joint; 2, inter-sesamoidean ligament; 3, 4, proximal end of digital synovial sheath; 5, ring formed by superficial flexor tendon; 6, fibrous tissue underlying ergot; 7, ergot; 8, 9, 9', branches of digital vessels; 10, distal ligament of distal sesamoid bone; 11, suspensory ligament of distal sesamoid bone; 12, 12', proximal and distal ends of bursa podotrochlearis. (From Sisson's "Anatomy of the Domestic Animals").

The first phalanx (os suffraginis) normally sets at an angle of about 50 to 55 degrees from a horizontal plane while weight is being supported. Its distal end articulates with the second or median phalanx (os corona) and forms the proximal interphalangeal (pastern or suffraginocoronary) joint. This also, is a ginglymus joint, having but slight lateral motion, and that only when it is in a state of flexion. A rather broad articular surface—from side to side—exists here, lessening the strain on the collateral ligaments somewhat. Dorsal flexion is checked by the flexor tendons and dorsal ligaments. Volar flexion is restrained by the extensor tendons.

The distal end of the second phalanx (os corona) has but slight lateral motion and this is manifested principally when it is in a state of volar flexion. Undue dorsal flexion is prevented by the deep flexor tendon (perforans) and volar flexion is inhibited by the extensor of the digit (extensor pedis). Thus it is seen, that when the leg is a weight-bearing member, weight is supported by the bony framework whose constituent parts are joined together by ligaments and tendons and each one of the several bones articulates in such manner that the joint is locked. The articular parts of bones rest upon or against an inhibitory apparatus, and are slightly flexed, as in the carpus, or considerably flexed such as in the fetlock joint when weight is being supported. In the first instance, for example, the flexors of the carpus and the superior check ligament assisted by the flexors of the phalanges constitute the inhibitory apparatus.

It will be noted that provision for weight bearing is so arranged that muscular energy is not required except in the matter of suspension of the body between the scapulae and here tonic impulses only are necessary to maintain an equilibrium[6], yet in every instance where weight is not supported by bones, inelastic ligaments or tendinous structures relieve the musculature of this constant strain. This explains the fact that some horses do not lie in the stall, yet in spite of their constant standing position, they are able to rest and sleep.

The student of lameness is interested in the function of the legs in the rôle of supporting weight and as propelling parts, and not particularly in the capacity of these members for inflicting offense or as weapons of defense. Yet, in the exercise of their functions other than that of locomotive appliances, injury often results, but usually it is the recipient of a blow that suffers the injury, such as an animal may

receive upon being kicked. Therefore, we do not often concern our-
selves with strains or other injuries that the subject experiences as
the result of efforts put forth in kicking or striking. Where such
injuries occur, however, a diagnosis is established by making use of
the principles heretofore discussed.

As propelling members the front legs bear weight and are advanced alternately when the horse is walking or trotting — in canter-
ing this is not so. When the normal subject travels in a straight line,
at a walk or a trot, the length of the stride is the same with the right
and left members. The stride of the right foot then, for example, is
equally divided by the imprint of the left foot, in the normal horse,
when traveling at a walk and in a straight line.

Shoulder Lameness.

This enigmatical term is frequently employed by the diagnostician when he is baffled in the matter of definitely locating the cause of lameness; when he has by exclusion and otherwise arrived at a decision that lameness is "high up." Shoulder lameness may be caused by any one or several of a number of conditions, e.g., fractures of the scapula or humerus; arthritis of the shoulder or elbow joint; luxation of the shoulder or elbow joint (rarely); injuries of muscles and tendons of the region due to strains, contusions or penetrant wounds; paralysis of the brachial plexus or of the prescapular nerve; involvement of lymph glands; arterial thrombosis; metastatic infections; rheumatic disturbances; and as the result of inflammation, infectious or non-infectious occasioned by collar bruises. In some instances such inflammation is due to the manner of treatment of collar injuries. Therefore, when one considers the numerous and dissimilar possible causes of shoulder lameness, it behooves the practitioner to become proficient in diagnostic principles.

A principle which is elemental in the diagnosis of locomotory impediment, is that lameness of the shoulder or hip is usually manifested by more or less difficulty in swinging the affected member. Swinging-leg-lameness, then, is usually present in shoulder affections. In some instances lameness is mixed as in joint ailments, involvement of the bicipital bursa (bursa intertubercularis), etc. In affections of the extremity there exists supporting leg lameness. Consequently, we employ this elemental principle, and, by a visual examination of the subject, which is being made to travel suitably, one may decide that lameness is either "high up"—shoulder lameness or, "low down"—of the extremity.

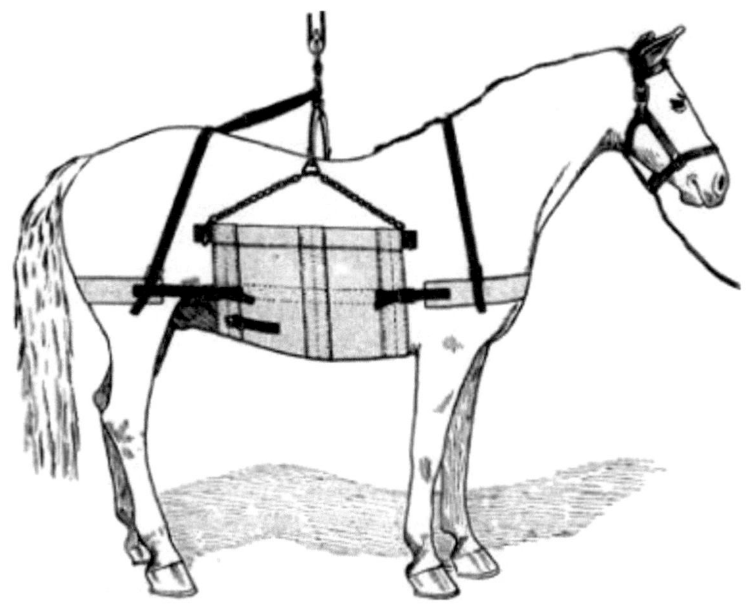

Fig. 5—Ordinary type of heavy sling.

To make practical use of this principle, the examiner must be thoroughly familiar with the anatomy of the various structures concerned in advancing the leg—those which support weight as well as those concerned both in weight bearing and swinging the member.

Fracture of the Scapula.

Etiology and Occurrence.—Fractures of the body of the scapula are of infrequent occurrence in horses for the reason that protection is afforded this bone because of its position. Its function, too, is such that very unusual conditions are necessary to subject it to fracture. The spine is occasionally broken due to blows such as kicks, etc., and here frequently a compound fracture exists.

Fig. 6—A sling made in two parts so that horses may be supported without use of central part or bodice. This sling is more comfortable than is the ordinary style and is particularly useful in cases that require a long period of this manner of confinement.

Where fractures of the body of the scapula occur, heavy contusions have been the cause as a rule, and serious injury is done the subject; consequently, treatment of fracture of the body of the scapula is seldom successfully practised. Fractures of the body of this bone resulting from accidents not involving internal injury or other disturbances and which would not seriously interfere with the vitality of the subject, are not necessarily serious unless compound.

Fractures of the neck of the scapula are serious because of the fact that there occurs displacement of the broken parts and perfect apposition of the fractured ends is difficult, if not impossible.

Fractures that extend to the articular surface are very serious, and complete recovery in such instances is practically impossible. The cartilage of prolongation of the scapula is sometimes seriously involved in certain cases of fistulous withers, and in some instances it has been separated from its attachment to the rhomboidea muscles, and lameness has resulted. In such instances, the upper portion of the scapula is disjoined from all attachment, and with every movement the animal makes, the scapula is moved back and forth. Complete recovery in such cases does not occur.

Symptomatology.—Fractures of the scapular spine are ordinarily readily recognized because there is usually visible displacement of the broken part. Crepitation is also detected without difficulty.

In fractures of the body of the scapula where an examination may be made before much swelling has taken place, and in subjects that are not heavily muscled, one should have no difficulty in recognizing the crepitation.

Fractures of the neck of the scapula are recognized by crepitation, by passively moving the leg, but it is necessary to exclude fractures of the humerus when one depends upon the finding of crepitation by this means. However, unless undue swelling exists, the exact location of the crepitation is recognized without serious difficulty.

Treatment.—The treatment of compound fractures of the scapular spine consists in the removal of the broken piece of bone by way of a cutaneous incision so situated that good drainage of the wound will follow.

Simple fractures of the body of the scapula are best treated by placing the subject in a sling, if the animal is halter broken, and enforcing absolute quiet for a period of from three to six weeks.

Splints or similar appliances are not of practical value in scapular fractures.

Compound fractures of the scapula usually result from violence, which at the same time does serious injury to adjacent structures, and it then becomes necessary to administer an expectant treatment, observing general surgical principles and providing in so far as possible for the comfort of the patient.

Scapulohumeral Arthritis.

Anatomy. — The scapulohumeral joint is an enarthrodial (ball and socket) joint wherein the ball or humeral articulating head greatly exceeds in size the socket or glenoid cavity of the scapula. The capsular ligament surrounding this joint is very large and admits of free and extensive movement of the articulation. There exist no lateral or common ligaments jointing the scapula and humerus as in other joints, but instead the tendinous portions of muscles perform this function. The principal ones which are attached to the scapula and humerus that act as ligaments are the supraspinatus (anteaspinatus), infraspinatus (postea-spinatus) biceps-brachii (flexor brachii) and subscapularis muscles.

Etiology and Occurrence. — Inflammation of the scapulohumeral articulation results from injuries of various kinds, including punctures which perforate the joint capsule, bruises from collars, metastatic infections and involvement as a result of direct extension of infectious conditions situated near the joint.

Classification. — Acute arthritis may be septic or aseptic, and there seems to be a remarkable tendency for recovery in cases of septic arthritis involving this joint in the horse.

Chronic arthritis with destruction of articular surfaces and ankylosis, is seldom observed. It is only in cases of severe injury, where the articular portions of the bones are damaged at the time of infliction of the injury, and where the articulation remains exposed for weeks at a time, together with immobility of the parts because of attending pain, that permanent ankylosis results.

Scapulohumeral arthritis may result then from *infections*, local or metastatic; from *injuries*, such as contusions of various kinds; from *wounds*, which break the surface structure or perforate the joint capsule; or from *luxations*.

Infectious Arthritis.

Infectious arthritis of the scapulohumeral joint the result of local causes other than produced by septic wounds, seldom causes serious inconvenience to the subject. Where such occurs, however, there is manifested mixed lameness and complete extension of the extremity is impossible. Local swelling is present and manifestations of pain are evident upon palpation of the affected area.

Treatment.—During the first stage of the infection, local applications, hot or cold, are indicated. A hot poultice of bran or other suitable material contained within a muslin sack, may be supported by means of cords or tapes which are passed over the withers and tied around the opposite fore leg. Such an appliance may be held in position more securely by attaching it to the affected member. Following the acute stage of such an infection, any local counter-irritating application or even a vesicant is in order.

Where abatement of the infectious process does not take place, and suppuration of the structures in the vicinity of the joint occurs, it is necessary to provide drainage for pus. In some cases of strangles, for instance, large pus cavities are formed and drainage is imperative. However, metastatic inflammation of this joint is seldom observed except in cases of strangles. The animal should be kept perfectly quiet until recovery has taken place.

Injuries.

Injuries to the scapulohumeral joint may be the result of kicks, runaway accidents or bruises from the collar, and there may result, because of such injuries, reactionary inflammation which will vary in intensity from the mildest synovitis to the most severe arthritis, causing more or less lameness.

Treatment.—The general plan of treatment in this form of arthritis is the same as has been outlined under the head of infectious arthritis, with the exception that there is seldom occasion to provide for drainage of pus.

Wounds.

Wounds which cause a break of the skin and fascia overlying the scapulohumeral joint are usually of little consequence, unless the blow is of sufficient force to directly injure the articulation, and in such cases, the treatment of the injury along general surgical principles, such as cleansing the area, providing drainage for wound secretion, and the administration of suitable dressing materials such as antiseptic dusting powder, is all that is required for the wound. The symptoms manifested by the subject in such cases are the same as have been discussed heretofore and merit no special consideration.

Prognosis.—Unless very serious injury be done the articular portions of the scapula or the humerus, resulting in the destruction of the capsular ligament, prognosis is entirely favorable.

Open Joint.—Where the capsular ligament is perforated and the condition becomes one of open joint, then a special wound treatment becomes necessary. The surface of the skin is first freed from all hair and filth in the vicinity of the wound. The wound proper is cleared of all foreign material either by clipping with the scissors, curetting or mopping with cotton or gauze pledgets. The whole exposed wound surface as well as the interior of the joint cavity, if much exposed, is moistened with tincture of iodin. Subsequent treatment consists in a local application of a desiccant dusting powder, which should be applied five or six times daily. The composition of the powder should be such as to permit of its liberal use, thereby affording mechanical protection to the wound as well as exerting a desiccative effect. Equal parts of boric acid and exsiccated alum serve very well in such cases.

Animals suffering from open joints of this kind should be confined in a standing position, preferably in slings, and kept so confined for three or four weeks. Since they usually bear weight upon the affected member, there is no danger of laminitis resulting.

Luxation of the Scapulohumeral Joint.

Because of the large humeral head articulating as it does with a glenoid cavity, scapulohumeral luxations are very rare in the horse. According to Moller[7], luxation is generally due to excessive flexion of the scapulohumeral joint. In such cases the head of the humerus is displaced anterior to the articular portion of the scapula and remains so fixed.

Symptoms.—Complete luxation of the scapula is recognized because of immobility of the scapulohumeral joint and of the abnormal position of the head of the humerus, which can be recognized by palpation, unless the swelling be excessive. Immobility of the scapulohumeral joint is noticeable when one attempts to passively move the parts.

Treatment.—Reduction of the luxation is effected by making use of the same general principles that are employed in the reduction of all luxations, and they are—the control of the animal so that the manipulations of the operator are not antagonized by muscular contraction, which is best accomplished by anesthesia; placing the luxated bones in the position which they have taken to become unjointed; and then making use of force which is directed in a manner opposite to that which has effected the luxation.

In a forward luxation of this kind, the operator should further flex the humerus, and while it is in this flexed position, force is exerted upon the articular head of this bone, and it is pushed downward and backward into its normal position.

After-care consists in restriction of exercise and, if necessary, confining the subject in a sling and the application of a vesicant over the scapulohumeral region.

Inflammation of the Bicipital Bursa.
(Bursitis Intertubercularis.)

Anatomy.—There is interposed between the tendon of the biceps brachii (flexor brachii) and the intertubercular or bicipital groove a heavy cartilaginous pad, which is a part of the bursa of the biceps brachii. This synovial bursa forms a smooth groove through which the biceps brachii glides in the anterior scapulohumeral region. Great strain is put upon these parts because the biceps brachii is the chief inhibiting structure of the scapulohumeral articulation—the one which prevents further flexion of the humerus during weight bearing. Passing, as it does, over two articulations, the biceps brachii has a somewhat complicated function, being a flexor of the radius and an extensor of the humerus. Thus it is seen, the biceps brachii is a weight bearing structure, as well as one that has to do with swinging the leg.

Etiology and Occurrence.—Because of the exposed position of the bicipital bursa (bursa-intertubercularis) it is occasionally injured. Blows and injuries received in runaway accidents do serious injury to the bursa and because of the peculiar and important part it plays during locomotion, serious injuries are not likely to resolve, and too often chronic lameness results. It is to be noted that the tendon of the biceps brachii (flexor brachii) is always involved in cases of inflammation of the bicipital bursa, and according to the late Dr. Bell[8] strain of the biceps brachii is a frequent cause of lameness in city horses, more frequent than is generally supposed.

Pathological Anatomy.—More or less destruction of the cartilaginous portion of the bursa, sometimes involving the tendinous portion of the biceps, takes place and, according to Moller, in some instances there occurs ossification of the tendon. Autopsies in some old horses reveal the presence of erosions of cartilage and hyperthrophy of the inflamed parts.

Symptoms.—In acute inflammations, there is always marked lameness. This is manifested to a greater degree when the subject advances the affected leg. There is incomplete advancement of the member; the toe is dragged when the horse is made to walk and the foot kept in a position posterior to the opposite or weight bearing foot while the subject is at rest. Lameness is disproportionate to the amount of local manifestation in the way of heat, swelling and pain that is to be recognized on palpation. In fact, in some cases so much pain attends the condition that no weight is borne by the affected member, and when compelled to walk, the subject hops on the sound leg.

Chronic inflammation of the bicipital bursa is occasionally met with wherein both members are affected. Because of the nature of the structures involved, when inflamed, chronic inflammation is a more frequent termination than is complete recovery. Bilateral affections are seen in horses that are driven for years, regularly at a fast pace on paved streets. In such cases, the gait is stilted, that is, there is incomplete advancement of both members and, of course, the period of weight bearing is correspondingly shortened; hence the short strides.

In chronic cases, little if any evidence of inflammation is to be detected by digital manipulation of the parts. If flinching occurs, one is often unable to interpret the manifestation as to whether it is due to inflammation or not.

There is no marked "warming out" in this condition, and animals are nearly as lame after having been driven a considerable distance as when started, although the lameness is not as a rule very great.

Treatment. — In very painful cases acute inflammation is treated by employing cold applications during the initial stage. Cracked ice when contained in a suitable sack may be held in contact with the affected part and the pack is supported by means of cords or tapes as suggested in the discussion on treatment of scapulohumeral arthritis on page 66. Later, hot applications may be employed to good advantage.

In the course of ten days or two weeks, if the acute painful condition has entirely subsided, vesication is indicated. The ordinary mercury and cantharides combination does very well. Depending upon the course taken in any given case, one is guided in the treatment employed. If prompt resolution comes to pass, the subject may be given free run at pasture after three or four weeks confinement in a box stall. If, however, the case does not progress in a prompt and satisfactory manner, absolute quiet must be enforced for six weeks or more. Repeated blistering is beneficial, although it is doubtful if firing is of sufficient benefit in the average chronic case of intertubercular bursitis to justify the punishment which this form of treatment inflicts, unless infliction of pain is the thing sought, to enforce repose in restless subjects. Patients are best given a long rest at pasture and returned to work for two or three months after an acute attack of inflammation of the bursa, lest the condition become chronic. When due consideration is given the pathology of such cases, the frequent unsatisfactory termination under the most careful treatment, is readily understood.

Contusions of the Triceps Brachii.
(Triceps Extensor Brachii: Caput Muscles.)

Anatomy.—The triceps brachii is the principal structure which fills the space between the posterior border of the scapula and the humerus. The several heads originate for the most part on the border of the scapula, the deltoid tuberosity of the humerus and the shaft of the humerus. Insertion of this large muscular mass is effected by means of several tendons to the olecranon. A synovial bursa is situated underneath the tendinous attachment of the posterior portion of the triceps brachii—the long head or caput magnum.

The function of the triceps as a whole is to flex the shoulder joint and extend the forearm. The triceps brachii is the chief antagonist of the biceps brachii.

Etiology and Occurrence.—Owing to the exposed position of this structure, it is not infrequently contused, the result of falls, kicks and other injuries. The function of the triceps is such that it becomes strained upon rare occasions when a horse resists confinement of restraint in such manner that the parts are unduly tensed in contraction. This sort of resistance may stretch the radial nerve or its branches in a way that paralysis results. A condition known as "dropped elbow" is described by Henry Taylor, F.R.C.V.S., in the Veterinary Record[9], wherein a two-year-old colt while resisting confinement was so injured.

The triceps group because of its convenient location, constitutes the site for hypodermic injection of drugs and biologic agents, with some practitioners; and as a result, more or less inflammation may occur. The author has observed and treated some twenty cases where an intensely painful infectious inflammation of the triceps brachii was caused by the intramuscular injection of a caustic solution by a cruel and unscrupulous empiric, whose object was to increase his practice.

Symptomatology.—As the triceps brachii is not particularly taxed during weight bearing in the subject at rest, there may be no unnatural position assumed during inflammation of the triceps. More or less swelling and supersensitiveness is always present, however, and great care and discrimination must be exercised in digital manipulation of the triceps region because many animals are normally sensitive to palpation of these parts. It is sometimes difficult to correctly interpret the true state of conditions because of this peculiarity.

There is always swinging-leg-lameness, which is accentuated when the subject is urged to trot. Where symptoms are pronounced, it is unnecessary to cause the subject to move at a faster pace than at a walk to recognize the condition. The forward stride is shortened and in extremley painful conditions, no attempt is made to extend the leg. It is simply carried *en une piéce*—flexion of the shoulder and elbow joints is carefully avoided.

Treatment.—During the early stage of inflammation, hot or cold applications are beneficial. Long continued use of moist heat—fomentations—allays pain and stimulates resolution. Keeping in contact with the painfully swollen parts a suitable bag filled with bran, which can be moistened at intervals with warm water, constitutes a practical and easy means of treatment. By employing this method, one is more likely to succeed in having his patient properly cared for, in that less work is entailed than if hot fomentations are prescribed.

After the acute and painful stage has subsided, a stimulating liniment is of benefit. The subject should be kept within a comfortable and roomy box stall for a sufficient length of time to favor prompt resolution. Wild and nervous subjects, if not so confined, will probably overexert the affected parts if allowed the freedom of a paddock or pasture.

Where the inflammation becomes infective, surgical interference is necessary. The prompt evacuation of pus, with adequate provision for wound discharge, should be attended to before extensive destruction of tissue takes place. Resolution is prompt as a rule in such cases because of the vascularity of the structures and the ease with which proper drainage may be effected. No special after-care is necessary if drainage is perfect, except that one should avoid injecting the wound cavity with aqueous solutions unless it be absolutely necessary to cleanse such cavity, and then it is best to swab the wound rather than to irrigate it freely.

Shoulder Atrophy.
(Sweeny or Swinney)

No satisfactory consideration of the pathogeny of this condition is recorded, but practitioners have long distinguished between muscular atrophies which are apparently caused without doing serious injury to nerves and muscular atrophy which seems to be due to nerve affection. In the first instance, recovery when proper attention is given, is prompt; whereas, in the latter, regeneration of the wasted tissues requires months in spite of the best sort of treatment.

The parts more frequently affected are the supra- and infrascapularis (antea- and posteaspinatus) muscles. But in some cases the triceps group is involved; however, this occurs in unusual and chronic affections. No doubt, these chronic cases are due to suspended innervation and are not to be classed with the ordinary case of atrophy of the abductor muscles of the humerus (supra- and infraspinatus) as in the usual case of "sweeny."

Occurrence.—Shoulder atrophy such as the general practitioner commonly meets with, is an affection, more often seen in young animals and it seems to be due to injuries of various kinds which contuse the muscles of the shoulder. Ill-fitting collars and pulling in a manner that there occurs side draft with unusual strain on the muscles of one side of the neck and shoulder, seem to be the more frequent causes of this trouble. Blows such as are occasioned by kicks and falls frequently result in atrophy of shoulder muscles.

Course.—In some cases a rapidly progressive atrophy characterizes the case and lameness and atrophy appear at about the same time. The affection in such instances does not recover spontaneously but constitutes a condition which requires prompt and rational treatment so that function may be fully restored to the parts involved.

Occasionally one may observe cases where there is but slight atrophy; where the disease progresses slowly and atrophy is not extensive or marked. In vigorous young animals that are left to run at pasture when so mildly affected, spontaneous recovery occurs.

Symptomatology.—Lameness is the first manifestation of shoulder atrophy, and in many cases where lameness is slight, the veterinarian may fail to discover the exact nature of the trouble if he is not very proficient as a diagnostician of lameness or if he is careless in taking into consideration obtainable history, age of the subject, etc. Because of the fact that the average layman believes that practically every case of fore-leg lameness wherein it is not obvious that the cause is elsewhere, is due to a shoulder affection of some kind, we may be too hasty in giving the client assurance that no "sweeny" exists. In some of these cases where a diagnosis of "shoulder lameness" has been made and the client has been assured that no sweeny exists, the patient is returned in about a week and there is then marked atrophy of one or both of the spinatus muscles.

A mixed type of lameness characterizes this affection, and in the average case there exists little evidence of local pain. The salient points in recognizing the condition are a consideration of history if obtainable; age of the subject; finding slight local soreness, by carefully manipulating the muscles which are usually involved; noting the character of the lameness if any is present; and where atrophy is evident, of course, the true condition is obvious.

Treatment.—Subcutaneous injections of equal parts of refined oil of turpentine and alcohol, with a suitable hypodermic syringe, is a practical and ordinarily effective treatment. From five to fifteen cubic centimeters (the quantity varies with the size of the animal), of this mixture is injected into the atrophied parts at different points, taking care to introduce only about one to two cubic centimeters at each point of injection. The syringe should be sterile and, needless to say, the site of injections must be surgically clean.

Other agents, such as tincture of iodin, solutions of silver nitrate, saline solutions and various more or less irritating preparations have been employed; but in the use of these preparations one may either fail to stimulate sufficient inflammation to cause regeneration to take place, or infection is apt to occur. Where suppuration results, surgical evacuation of pus must be promptly effected else large suppurating cavities form.

The employment of setons constitutes a dependable method of treatment of shoulder atrophy, but because of the attendant suppurative process which inevitably results, this method is not popular with modern surgeons and is a last resort procedure.

After-care.—Regular exercise such as the horse usually takes when at pasture, is very helpful in treating atrophy, and in some cases it has been found that no reasonable amount of irritation would stimulate muscular regeneration; but by later allowing patients to exercise at will, recovery took place in a satisfactory manner. No special attention is ordinarily necessary.

Paralysis of the Suprascapular Nerve.

Anatomy. — The suprascapular (anterior scapular) nerve, a small branch of the brachial plexus, is given off from the anterior portion of this plexus. The nerve rounds the anterior border of the neck of the scapula, passing upward and backward under the supraspinatus (antea-spinatus) muscle and terminating in the infraspinatus (postea-spinatus) muscle.

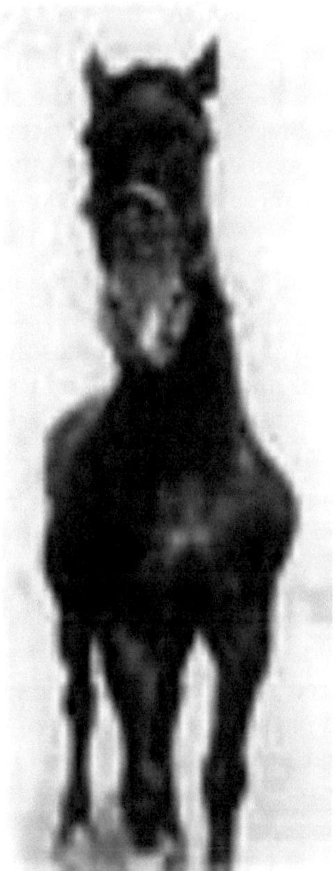

Fig. 7 — Paralysis of the
suprascapular nerve
of the left shoulder

Etiology and Occurrence.—As the result of direct injury to this nerve by contusion such as may be received in runaway accidents, collar bruises, especially collar bruises in young horses that are not accustomed to pulling and that walk in a manner to cause side draft, injury to the nerve occurs, and partial or complete paralysis supervenes. Some writers state that it may be produced by confining an animal in recumbency, with the casting harness. The common cause of paralysis or paresis of this nerve in cases such as one observes in country practice, is bruises from the collar in colts that are put to heavy farm work or where ill fitting collars are used.

Symptomatology.—With partial or complete suspension of function of the suprascapular nerve there results enervation of the supraspinatus and infraspinatus muscles. Since these muscles act as external lateral ligaments of the scapulohumeral joint, when they are incapacitated, there naturally follows more or less abduction of the shoulder when weight is borne.

In extreme cases, as soon as the ailing animal is caused to support weight with the affected member, the joint is suddenly thrown outward in a manner that the average layman at once concludes that there must be scapulohumeral luxation, and the veterinarian receives a call to see a case wherein the "shoulder is out of place." There exists, however, no luxation in such cases.

If serious injury is done the nerve so that it undergoes degenerative changes, there will result atrophy of the muscles that derive their nerve supply from the suprascapular nerve.

Treatment.—During the first few days following injuries which result in this form of paralysis, it is well to keep the subject inactive, and if much inflammation of the injured structures contiguous to the nerve exists, the application of cold packs is beneficial. Later, as soon as acute inflammation has subsided, vesication of a liberal area around the anteroexternal part of the scapulohumeral joint and over the course of the suprascapular nerve, will stimulate recovery in favorable cases. As a rule, in mild cases, the subject is in a condition to return to work in two or three weeks.

Radial Paralysis.

Described under the titles of "Radial Paralysis" and "Brachial Paralysis," there is to be found in veterinary literature a discussion of conditions which vary in character from the almost insignificant form of paresis to the incurably affected conditions wherein the whole shoulder is completely paralyzed.

When one considers the anatomy of the brachial nerve plexus and the distribution of its various branches, the location of this plexus and its proximity to the first rib, and the inevitable injury it must suffer in fracture of this bone, together with the inaccessibility of the plexus, it is not strange that a correct diagnosis of the various affections of the brachial plexus and the radial nerve is often impossible until several days or weeks have passed. And, in some instances, diagnosis is not established until an autopsy has been performed. Here, too, we fail to find cause for paralysis in some rare instances.

Anatomy.—The radial nerve is a large branch of the brachial plexus and is chiefly derived from the first thoracic root of the plexus and is here situated posterior to the deep brachial artery. It is directed downward and backward under the subscapularis and teres major muscles, rounding the posterior part of the humerus, and passing to the anterior and distal end of the humerus, it finally terminates in the anterior carpal region. The radial nerve supplies branches to the three heads of the triceps brachii, to the common and lateral extensors of the digit and also to the skin covering the forearm.

Etiology and Occurrence.—Nothing definite is known about the cause of some forms of radial paralysis. However, radial paralysis is encountered following injury to the nerve occasioned by its being stretched, as in cases where the triceps brachii is unduly extended in restraining subjects by means of a casting harness. Berns[10] states that in confining horses on an old operating table where it was necessary to draw the affected foot forward twenty-four to thirty-six inches in advance of its fellow, which was secured in a natural vertical position, radial paralysis of a mild form was of frequent occurrence. Country practitioners, in restraining colts by casting with harness or ropes, occasionally observe a form of paresis wherein the radial nerve suffers sufficient injury that there is caused a temporary loss of function of the triceps brachii. Such cases recover within three or four days and are not a true paralysis, but nevertheless constitute conditions wherein normal nerve function is temporarily suspended.

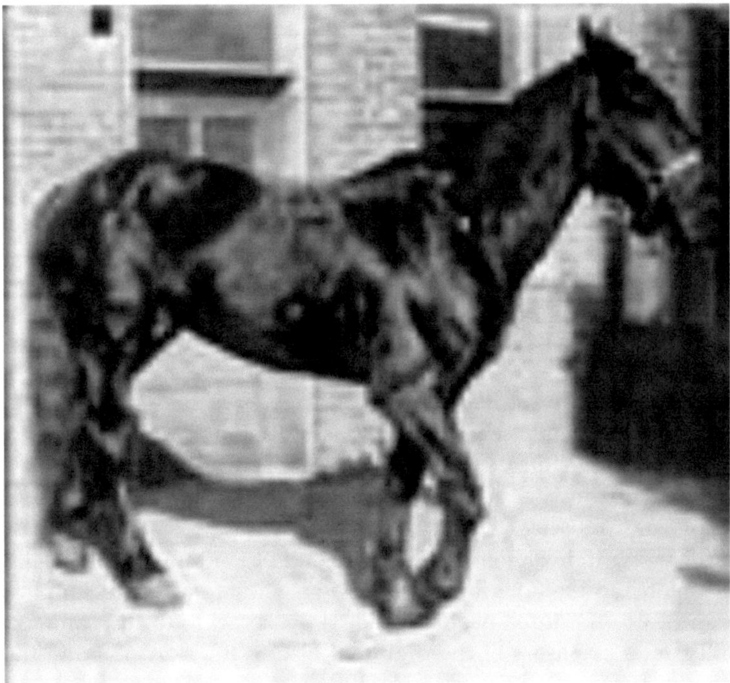

Fig. 8—Radial paralysis.

Symptoms.—Immediately subsequent to injuries which involve the radial nerve, there is manifested more or less impairment of function. Remembering the structures supplied by the radial nerve and its branches, one can readily understand that there should occur as Cadiot[11] has stated:

In complete paralysis, the joints of the affected limb with the exception of the shoulder are usually flexed when the horse is resting. In consequence of loss of power in the triceps and anterior brachial muscles, the arm is extended and straightened on the shoulder, the scapulohumeral angle is open, and the elbow depressed. The fore-

114

arm is flexed on the arm by the contraction of the coracoradialis (biceps brachii), while the metacarpus and phalanges are bent by the action of the posterior antibrachial muscles. The knee is carried in advance, level with, or in front of, a vertical line dropped from the point of the shoulder. The hoof is usually rested on the toe, but when advanced beyond the above mentioned vertical line, it may be placed flat on the ground, the joints then being less markedly bent. When the limb as a whole is flexed, it may be brought into normal position by thrusting back the knee with sufficient force to counter-act the action of the flexor muscles.

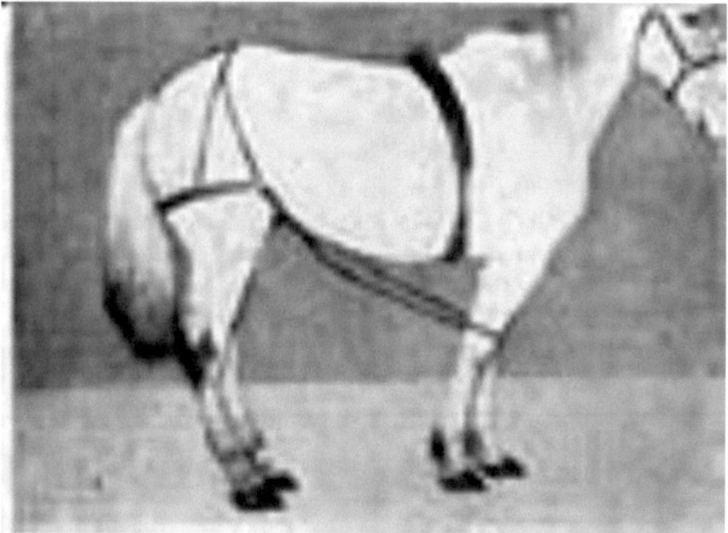

Fig. 9—Merillat's method of fixing carpus in radial paralysis. Courtesy, Alex. Eger.

When made to walk, the animal being unable to exert muscular action with the paralyzed structures, limply carries the member as a whole, and there is shortening of the anterior portion of the stride. There being loss of function of the triceps brachii, it is impossible for the subject to straighten the leg in the normal position for supporting weight; therefore, any attempt to bear weight results in further flexion of the affected member and the animal will fall if the body is not suddenly caught up with the sound leg.

Differential Diagnosis.—In making examination of these cases, one can exclude fracture by absence of crepitation and usually, also, swelling is absent in radial paralysis. In a typical case of radial paralysis, the affected leg can sustain its normal share of weight if placed in position, that is, if the carpal joint is extended in such manner that the leg is positioned as in its normal weight-bearing attitude. In brachial paralysis, whether due to fracture of the first rib or to other serious injury, it is impossible for the subject to support weight with the affected member even when it is passively placed in position.

No difficulty is ordinarily experienced in differentiating radial paralysis from muscular injuries to the triceps; yet, in some cases of "dropped elbow," it is necessary to observe the progress of the case for ten days or two weeks before one can positively establish a diagnosis.

Quoting Merillat[12]: "When, after four weeks, there is no amelioration of the paralysis, the muscles have atrophied, and the patient has become emaciated from pain and discomfort, the diagnosis of brachial paralysis with fracture of the first rib may then be announced."

Prognosis.—When no complete paralysis of the brachial plexus or no fracture of the first rib exists, the majority of cases recover completely in from ten days to six weeks. Some writers claim that recoveries occur in ninety per cent of cases when conditions are favorable.

Treatment.—When incomplete radial paralysis exists, little needs be done except to allow the subject moderate exercise and to provide for its comfort. Local applications, stimulative in character, are beneficial, and the internal administration of strychnin is indicated.

In the cases where weight is not supported without the affected leg being passively placed in position, it is necessary to provide for the subject's comfort in several ways.

Mechanical appliances such as braces of some kind in order to keep the affected leg in a position of carpal extension, constitute the essential part of treatment. The leg is supported in such a manner that flexion of the carpus is impossible. Due regard is given to prevent chafing or pressure necrosis by contact of the skin with the braces—this may be done by bandaging with cotton. The supportive appliance is kept in position for ten days or two weeks. At the end of this time the brace may be removed and the subject given a chance to walk, and improvement, if any exists, will be evident. When there is manifested an amelioration of the condition, moderate daily exercise and massage of the affected parts are helpful.

Should the subject be seriously inconvenienced by the application of a brace or other supportive appliances, it is necessary to employ slings. Further, if weight is supported entirely by the unaffected member, laminitis may supervene if a sling is not used.

Thrombosis of the Brachial Artery.

Thrombosis of the brachial artery or of its principal branches is of very rare occurrence in horses.

Etiology.—Partial or complete obstruction of arteries (brachial or others) occurs as the result of direct injury to the vessel wall from compression and tension of muscles and resultant arteritis; lodging of emboli; and parasitic invasion of vessel walls causing internal arteritis.

Symptomatology.—If sufficient collateral circulation exists to supply the parts with blood, no inconvenience is manifested while the subject is at rest. Where the lumen of the affected vessel is not completely occluded, there may be no manifestation of lameness when the ailing animal is moderately exercised. Consequently, the degree of lameness depends upon the extent of the obstruction to circulation; and, likewise, the course and prognosis depend upon the character and extent of such obstruction.

In severe cases, lameness is markedly increased by causing the animal to travel at a fast pace for only a short distance. There are evinced symptoms of pain, muscular tremors and sudation, but the affected member remains dry and there is a marked difference of temperature between the normal areas and the cool anemic parts. When the subject is allowed to rest, circulation is not taxed, and there is a return to the original and apparently normal condition, only to recur again with exertion. This condition characterizes thrombosis.

Treatment.—In these cases, little if any good directly results from any sort of treatment in the way of medication. Absolute rest is thought to be helpful. Potassium iodid, alkaline agents such as ammonium carbonate and potassium carbonate, have been administered. Circulatory stimulants also have been given, but it is doubtful if any good has come from medication.

Fracture of Humerus.

The shaft of the humerus, protected as it is by heavy muscles, is not frequently fractured; and fractures of its less protected parts, as for example, the head, are complicated in such manner that resultant arthritis soon constitutes the more serious condition.

As a result of falls on frozen ground, kicks or any other form of heavy contusion, the humerus is occasionally broken. It is rarely fractured otherwise. Because of the force of contusions usually required to effect humeral fracture, the manner in which the bone is broken, with respect to direction, is variable. Often oblique fractures exist and occasionally there occurs multiple fracture. In addition to the ordinarily serious nature of the fracture itself, there is always much injury done the adjoining structures.

Symptomatology.—Mixed lameness and manifestation of severe pain characterize this affection. Considerable swelling which increases, in some cases for a week or more, is to be observed. Crepitation is readily detected, if pain and swelling is not too great to prevent passive movement of the member. Where intense pain is not manifested, because of manipulation, one may abduct the extremity and thereby occasion distinct crepitation; but when it is possible to recognize crepitation by holding the hand in contact with the olecranon while the animal is made to walk, this method is to be preferred, if the subject can move without serious difficulty. The pathognomonic symptom here is recognition of crepitation, but this may be very difficult to recognize in fracture of condyles, and in such instances, a careful examination is necessary. Gentle manipulation in a manner that pain is not aggravated will tend to inspire confidence on the part of the subject and relaxation of muscles will enable the operator to detect crepitation.

Course and Prognosis.—Because of the direction of the long axis of the humerus, with relation to the bony column of the extremity, it is obvious that any lateral movement of the leg tends to rotate the shaft of this bone. In fractures of the shaft of the humerus, then, it is apparent that immobilization is very difficult if at all possible.

The proximity to the axillary lymph glands makes for easy dissemination of infection when the contused musculature becomes infected. The adjacent brachial nerve plexus is so very apt to become involved, if not actually injured at the time fracture occurs, that paralysis is a probable complication. Consequently, it is logical to reason that because of the many possible serious complications, such as shock, occasioned by the injury and the distress and pain which this accident produces, recovery must be the exception in fracture of the humerus. However, recoveries do take place and in addition to the reported recoveries by Liautard, Moller, Stockfleth, Lafosse, Frohner and others, we have instances cited by American practitioners where cases resulted in recovery. Thompson[13] reports a good recovery in a 1600-pound mare where there existed an oblique fracture of the humerus. This mare was kept in slings for eight weeks. Walters[14] reports complete recovery in humeral fracture in a foal three days old. The only treatment given was the application of a pitch plaster from the top of the scapula to the radius. The colt was kept in a comfortable box stall and in about four weeks regained use of the leg. Complete recovery eventually resulted. In the experience of the author, recovery has not occurred in humeral fractures.

Treatment.—When animals are not aged and of sufficient value to justify treatment, they are best supported in a sling, if halter broken. If subjects are nervous, wild and unbroken, it is possible to employ the sling, if care is given to train the animal to this manner of restraint. The presence of an attendant for a day or two will reassure such subjects so that even in these cases it may be practicable to employ the sling.

Braces and other mechanical appliances intended to immobilize the parts are not of practical benefit in the horse. Unlike the dog, the horse as yet has not been successfully subjected to tolerating rigid braces for the shoulder and hip.

Everything possible must be done that will make for the patient's comfort. If the subject turns out to be a good self nurse, and the nature of the fracture is such that practical apposition of the broken ends of bone may be maintained, recovery will occur in some cases.

Inflammation of the Elbow.
(Arthritis.)

Affections of this articulation other than those which are produced by traumatism are rare. This joint has wide articular surfaces, and securely joined as they are by the heavy medial and lateral ligaments (internal and external lateral ligaments), luxation is practically impossible. When luxation does occur, irreparable injury is usually done. Castagné as quoted by Liautard[15], reports a case of true luxation of the elbow joint in a horse where reduction was effected and complete recovery took place at the end of twenty-five days. This is an unusual case. The average practitioner does not meet with such instances.

Anatomy. — The condyles of the humerus articulate with the glenoid cavities of the radius and a portion of the ulna. Two strong collateral ligaments pass from the distal end of the humerus to the head of the radius. The capsular ligament is a large, loose membrane which encloses the articular portion of the humerus with the radius and ulna and also the radioulnar articulation. It is attached anteriorly to the tendon of the biceps brachii (flexor brachii). The capsule extends downward beneath the origin of these digital flexors. This fact should be remembered in dealing with puncture wounds in the region, lest an error be made in estimating their extent and an open joint be overlooked at the initial examination.

Etiology and Occurrence. — Exclusive of specific or metastatic arthritis, which is seldom observed except in young animals, inflammation of the elbow joint is usually caused by injury. This articulation is not subject to pathologic changes due to concussion or sprains as occasioned by ordinary service, but is frequently injured by contusion from falls, blows from the wagon-pole and kicks. Wounds which affect the elbow joint, then, may be thought of in most cases, as resultant from external violence. They may be contused wounds or penetrant wounds. Sharp shoe-calks afford a means of infliction of penetrant wounds which may occasion open joint and infectious arthritis.

Classification.—A practical manner of classifying inflammation of the elbow is on an etiological basis. Eliminating the forms of elbow inflammation, such as are caused by metastatic infection and other conditions which properly belong to the domain of theory of practice, we may consider this affection under the classification of *contusive wounds* and *penetrative wounds*.

Symptomatology.—Any injury which is of sufficient violence to occasion inflammation of the elbow causes marked lameness and manifestation of pain. The degree of lameness and distress manifested by the subject, depends upon the nature and extent of the involvement. A contusion suffered as the result of a fall, which occasions a circumscribed inflammation of the structures covering this joint and where little inflammation of the articulating parts exists, marked evidence of pain and lameness might be absent. On the other hand, if a true arthritis is incited, there will be evident distress manifested, such as hurried respiration, accelerated pulse, inappetence, mixed lameness, local evidence of inflammation and particularly marked supersensitiveness of the affected parts. Considering these two extremes of manifested distress and injury, one may readily conclude that in the frequently seen case, wherein contusion has occasioned a moderate degree of injury, prognosis is favorable and recovery ordinarily follows in the course of a few weeks' treatment.

In cases of arthritis due to penetrative wounds (because of the important function of this joint and its large capsule, which when inflamed discharges synovia in a manner that closure of such an open joint is seldom possible) a very grave condition results.

Treatment.—Inflammation of the elbow, such as is frequently seen in general practice where horses are turned out together and exposed to kicks and other injuries, yields to treatment readily, if an open joint does not exist.

Hot packs supported in contact with the elbow and kept around the inflamed articulation for a few days, materially decrease pain and tend to reduce inflammation. The subject must be kept quiet in a comfortable stall and, if necessary, a sling used. Where it is impossible for the animal to support much weight with the injured member the sling should be employed.

As inflammation abates, which it does in the course of from one to three weeks in uncomplicated cases, the subject may be allowed the freedom of a comfortable box stall. Vesication of the parts is in order, and this may be repeated in the course of two weeks, if it is deemed necessary.

Penetrative wounds resulting in open joint are not treated with success as a rule, and because of the handicap under which veterinarians labor, methods of handling such cases, where large, important articulations are affected, are not being rapidly improved. Prognosis is usually unfavorable, and for humane and economic reasons, animals so affected should be destroyed.

Ordinary wounds of the region of the elbow are treated along general lines usually employed. They merit no special consideration, except that it may be mentioned that with such injuries concomitant contusion of the parts occasions injury that does not recover quickly.

Fracture of the Ulna.

Etiology and Occurrence.—Fractures of the ulna in the horse are not common in spite of the exposed position of the olecranon. This bone when broken, is usually fractured by heavy blows and any form of ulnar fracture is serious because of its function and position in relation to the joint capsule. Transverse fractures do not readily unite because of the tension of the triceps muscles, which prevent close approximation of the broken ends of the bone.

Thompson[16], however, reports a case of transverse simple fracture of the ulna in a mare, the result of a kick, in which complete recovery took place. He kept the subject in a sling for six weeks and then allowed six months rest.

Symptomatology.—The position assumed by a horse suffering from a transverse fracture of the ulna, is similar to that in radial paralysis. Crepitation may be detected by manipulating the parts, and in some instances of fracture of the olecranon, there occurs marked displacement of the broken portions of the bone. Lameness is intense and the parts are swollen and supersensitive. The capsular ligament of the elbow joint is usually involved in the injury because fracture of the ulna may directly extend within the capsular ligament. In such cases, there is synovitis, and later arthritis causes a fatal termination.

Treatment.—The impossibility of applying a bandage in any way to practically immobilize these parts in fracture of the ulna, prevents our employing bandages and splints. Therefore, one can do little else than to put the patient in a sling and try to keep it quiet and as nearly comfortable as circumstances allow.

Fracture of the Radius.

Etiology and Occurrence.—From heavy blows received such as kicks, collision with trees or in falls in runaway accidents, the radius is occasionally fractured. In very young foals, fracture of the radius, as well as of the tibia and other bones, results from their being trampled upon by the mother.

Symptomatology.—Excepting in some cases of radial fracture of foals where considerable swelling has taken place, there is no difficulty in readily recognizing this condition. The heavy brachial fascia materially contributes to the support of the radius, and in cases where swelling is marked, crepitation may not be readily detected. In fact, a sub-periosteal fracture may exist for several days or a week or more and then, with subsequent fracture of the periosteum, crepitation and abnormal mobility of the member are to be recognized. In such cases, the subject will bear some weight upon the affected member, but this causes much distress. In one instance the author observed a transverse fracture of the lower third of the radius which was not positively diagnosed until about ten days after injury was inflicted. In this case, without doubt, the subject originally suffered a sub-periosteal fracture of the bone and because the animal was a good self nurse, the brachial fascia supported the radius until the periosteum gave way and the leg dangled. In this instance infection took place and suppuration resulted. It was deemed advisable to destroy this animal.

Prognosis.—In adult animals, radial fracture constitutes a grave condition; generally speaking, prognosis, in such cases, is unfavorable. Because of the leverage afforded by the extremity, immobilization of the radius is difficult. Any sort of mechanical appliance, which will immobilize these parts, is likely to produce pressure-necrosis of the soft structures so contacted. There is occasioned thereby much pain and the subject becomes restive, unmanageable and sometimes the splints are completely deranged because of the animal's struggles, and much additional injury to the leg is done. Occasionally, an otherwise favorable case is thus rendered hopelessly impossible to handle, and the subject must be destroyed several days after treatment has been instituted.

Consequently, unless all conditions are good, and the affected animal a favorable subject, young, of good disposition, and the fracture a simple transverse one, complete recovery is not likely to result from any practical means of handling.

Treatment.—Mature subjects ought to be put in slings and kept so restrained throughout the entire time of treatment. Immobilization of the broken parts of the bone is the object sought. This is attempted by practitioners who employ various methods, and each method has its advocates.

Casts are used by some and serve very well in many cases; but because of their bulk and unyielding and rigid nature, they are not well adapted to use on fractures of bones proximal to the carpus and tarsus. This is in reference to plaster-of-paris casts or those of any similar material.

Appliances which depend on glue or other adhesive substances combined with leather, wood or fiber for their support, are efficacious but not comfortable.

The use of heavy leather when the member has been suitably padded with cotton and bandages, constitutes a very good manner of reducing fracture of the radius or of the tibia. Leather when cut to fit both the medial and lateral sides of a leg, and firmly held with bandages, will form a firm support that yields slightly to changes of position, thus making for comfort of the subject.

Such a splint or support should extend from the fetlock region to the elbow, but the cotton and bandages are to reach to the foot. When one considers that, with the supportive appliance placed on each side of the affected member, rigidity is accomplished as much from tensile strain put upon the leather as from its own stiffness, it is seen that the leather need not be of the heaviest—sole leather is unnecessary. Because of the more comfortable immobilizing appliance, the subject is less restive, and chances for a successful outcome are materially increased thereby.

In the mature subject, six or eight weeks' time is required for union of the parts to occur sufficiently so that splints may be dispensed with. Rearrangement of the supportive apparatus, however, is possible and usually necessary during the first few weeks of treatment. By employing care in handling the parts, the subject will be unlikely to do itself injury at the time readjustment of splints is being effected.

In foals, it is best to give them the run of a box stall with the mother. Being agile, they get up and lie at will without doing injury to the fractured member. The splints (leather is preferable in these cases also) are looked after and readjusted as necessity demands.

Three or four weeks time is all that is required for the average young colt to be kept in splints when suffering from simple transverse fracture of the radius.

Compound fractures are necessarily more difficult to treat than are the simple variety, but even in such cases recovery results sometimes, and the practitioner is justified in attempting treatment after having explained the situation to his client.

Oblique fractures, even when simple, do not completely recover. Muscular and tendinous contraction, together with the natural tendency for the beveled contacting parts of the broken bone to pass one another in oblique fracture, results in shortening of the leg and, if union results, a large callus usually forms. Where shortening of bones occur, necessarily, permanent lameness follows.

Wounds of the Anterior Brachial Region.

Etiology and Occurrence. — Contusions and lacerations of the forearm are of frequent occurrence in horses and are troublesome cases to handle; particularly is this noticeable where extensive laceration of the parts occurs. These injuries are caused by animals being kicked; by striking the forearm against bars in jumping; and in sections of the country where barbed wire is used to enclose pastures, extensive lacerated wounds are met with when horses jump into such fences.

Symptomatology. — Any wound which causes inflammation of the structures of the anterior half of the forearm, is characterized by swinging-leg-lameness. Depending upon the nature and extent of the injury, manifestation varies. In cases where laceration has practically divided all of the substance of the extensor tendons, it is, of course, impossible for the subject to advance the leg; but where lacerated wounds involve only a part of the extensor apparatus of the foreleg, not so much inconvenience is evident, unless the wound is seriously infected and inflammation involves contiguous structures. Therefore, in many instances, lameness is more pronounced in contusions of the anterior brachial region than where tissues have been divided more or less keenly.

In every instance diagnosis is easily established. The injury is quite evident, and the manner of locomotion is not in itself an essential feature to be considered in a discussion of symptoms. Where a contusion of the anterior brachial structures occurs, there is, in addition to lameness, swelling which is painful because of the pressure occasioned by the heavy non-yielding brachial fascia. And where suppuration occurs, there is then an intensely painful condition which is not relieved until pus has been evacuated. Rather frequently, drainage for wound secretions is a difficult problem, and approximation of the divided ends of muscles is always difficult to maintain.

Treatment. — Contused wounds of the anterior brachial region are treated along usual lines; that is, attempt is made to stimulate prompt resolution. Hot or cold applications are employed throughout the acute stage of the affection. Complete rest is provided for

until all pain has subsided. Later, stimulating liniments are beneficial.

Where no injury is done the periosteum or bone, complete resorption of all products of inflammation usually occurs, though in many instances, this is tardy — six weeks or more are sometimes required for recovery to take place.

If suppuration occurs, it is necessary to provide for drainage as soon as it is possible to distinguish the presence of pus. Due regard is given the manner of establishing drainage because of the usual existence of sub-fascial fistulae. In these cases, one avoids injecting solutions of aqueous antiseptics. By gently compressing the parts, pus is caused to drain out and in enforcing a moderate amount of exercise at a walk, when lameness is not intense, drainage is maintained. Cotton packs, moistened with hot antiseptic solutions, and kept around the forearm for several hours daily, are helpful because drainage is facilitated, and resolution is stimulated by the increase of blood thus attracted to the parts, and pain materially diminishes.

In lacerated wounds of the anterior brachial region, after having controlled hemorrhage, an area around the wound margin is freed of hair by clipping or shaving. The wound is carefully examined, and the best site for drainage is selected and a suitable opening for wound discharge is provided for. Where the extensor carpiradialis (metacarpi magnus) with other structures, is divided and the distal portion is torn downward, as frequently is the case in barbed wire cuts, it is necessary to make careful provision for drainage. The wound is thoroughly cleansed by means of ablutions if necessary; but preferably by swabbing with pledgets of cotton or gauze which are moistened in antiseptic solutions. All shreds of macerated tissue are clipped with scissors and finally the whole wound surface is painted with tincture of iodin.

If drainage is made by cutting through the tissues in the median portion of the structures that have been displaced, the opening should be packed with gauze so that it may remain patent after swelling has occurred. Such packing is left *in situ* for twenty-four hours.

The pendant muscular portions of tissues are sutured up by means of tapes and, while perfect apposition is not ordinarily possible, it is very essential to train the pendant tissues in their normal position even if they require resuturing within a week. This minimizes granulation of tissue, and there results less scar if the detached portions are kept near, even if not in contact with the proximal wound margins. The skin together with subcutaneous fascia is sutured on either side unless drainage is to be provided for on one side, and the lowermost part of that side is left unsutured.

After-care. — Where extensive suturing of tissues has been necessary, subjects must be kept quiet. They are best confined in box stalls and not taken out for several weeks. Particularly is this true where transverse division of extensors has taken place. Sutures are removed at the end of from ten days to three weeks as cases permit. Drainage of wound secretions, which usually become infected, is necessary, because with obstructed drainage in an infected wound of this kind, there will result an early destruction of tissue at some point sutured. Daily irrigation done in a manner that practical asepsis is carried out, is necessary for about a week. All irrigation is done by way of the drainage opening, and this with warm aqueous solutions of suitable antiseptics. After a week or ten days' time, the wound should not be dressed more frequently than twice weekly.

If it is necessary to leave a portion of the wound uncovered, as in cases where skin is destroyed, the frequent (three or four daily) application of a suitable antiseptic powder is necessary to check exuberant granulation. This may be directly effected by the use of an astringent or desiccant preparation, and such dressing serves as a mechanical protection as well.

When such wounds are kept clean, where drainage is properly maintained, and the subject kept quiet, no particular attention other than the local application of an astringent lotion (such as the zinc and lead lotion) is necessary after the first three or four weeks. Usually, if the animal gnaws at the parts or otherwise manifests evidence of discomfort, it is an indication that new areas of infection are being established because of obstructed drainage or retained eschars. A thorough cleansing of the wound with a two per cent solution of Liquor Cresolis Compositus and this followed by moistening every part of the wound with tincture of iodin, will check all such disturbance if done promptly.

Where practically all of the anterior surface of the radius has been denuded, recovery is tardy and there is in some cases imperfect extension of the leg for months after the wound has healed. But in such instances, animals gradually regain complete use of the affected member and in the course of a year function is fully restored.

Inflammation and Contraction of the Carpal Flexors.

Anatomy. — The structures which are usually considered as true flexors of the carpus are a group of three muscles, which have separate heads of origin and different points of tendinous insertion.

The *flexor carpiradialis* (flexor metacarpi internus) originates from the medial epicondyle of the humerus. It is inserted to the proximal end of the medial metacarpal (inner metacarpal or splint) bone. This muscle is the smaller of the three and is not usually divided in doing carpal tenotomy.

The *flexor carpiulnaris* (flexor metacarpi medius) has two heads of origin; one, the larger, originates from the epicondyle of the humerus and the other from the posterior surface of the olecranon. The two heads unite at the upper third of the radius and the muscle, becoming tendinous, as is the case with the other carpal flexors, is attached by one point of insertion to the accessory carpal bone (trapezum). The other blends with the posterior annular ligament of the carpus.

The *ulnaris lateralis* (flexor metacarpi externus) has its origin from the lateral epicondyle of the humerus and inserts to the proximal extremity of the fourth metacarpal (outer splint) bone and by another attachment to the accessory carpal bone (trapezium) with the tendon of the flexor carpiulnaris (flexor metacarpi medius).

Acting together, these muscles flex the carpus or extend the elbow and this action is antagonized by the biceps brachii (flexor brachii) and extensors of the carpus and phalanges.

Etiology and Occurrence. — Inflammation of the muscular or tendinous parts of the carpal flexors, does not occur as frequently as does inflammation of the flexors of the extremity. They are subject to injury such as is occasioned by hard work and concussion and contract as a result; but, more frequently a congenital malformation of the leg is responsible for undue strain upon these parts. Horses that are "knee sprung" or that have a congenital condition where in the anterior line, as formed by the radius, carpal and metacarpal bones, is bent forward at the carpus, are subject to inflammation and contraction of the carpal flexors. When these flexors are contracted, the condition is commonly known among horsemen as "buck knee." In itself, inflammation of the carpal flexors is not a condition which is likely to prove troublesome, but because of carpal involvement (which is often present) the cause of the trouble remains, and inflammation of the carpal flexors recurs or becomes chronic and contraction of tendons results.

Symptomatology.—Inflammation of the carpal flexors, when acute and uncomplicated, is characterized by a painfully swollen condition of the affected tendons. No weight is borne upon the affected leg and the carpal joint is flexed. Mixed lameness is present. There is no difficulty encountered in arriving at a diagnosis because of the very noticeably inflamed parts.

Many fully developed cases of contraction of the tendons of the carpal flexors are observed where the condition has become established gradually and no lameness has resulted from tendinitis or carpitis. In some of these cases, subjects are stumblers and when they are carelessly handled or kept at fast work over irregular or hard roads, chronic carpitis with hyperplasia of the structures of the anterior carpal region results, owing to frequent bruising from falls.

Fig. 10—Contraction of carpal flexors, "knee sprung."

Where inflammation is caused by a puncture wound and subfascial infection occurs, there is evident manifestation of pain. No weight is supported by the affected member and because of the pressure, occasioned by the swollen muscles confined within the non-yielding brachial fascia, there exists marked supersensitiveness of the affected parts. Flexion of the elbow is avoided because contraction of the biceps brachii (flexor brachii) or the extensors, which are antagonists of the flexors of the carpus, tenses the carpal flexors and pain is thereby increased.

However, in most instances, the practitioner's attention is not directed to typical and uncomplicated cases, but to subacute or chronic inflammations which are often attended with contraction of the tendinous parts of the carpal flexors, and in such cases carpitis is present. Animals so affected have lost the rigidity which characterizes the normal carpal joint when the leg is a weight bearing member, and because of its sprung condition, the leg trembles when supporting weight.

Treatment. — Acute inflammation is treated by means of local application of cold or hot packs until the pain and acute stage of inflammation has subsided and later stimulating liniments are indicated. Absolute quiet must be enforced. Especially where the carpus is involved must the subject be kept quiet until all evidence of inflammation has subsided.

The application of vesicants or line-firing is beneficial in subacute inflammation of the tendons of the carpal flexors. Where contraction of tendons exists and no osseous or ligamentous change prevents correction of the condition, tenotomy is necessary. The reader is referred to Merillat's "Veterinary Surgery"[17] for a good description of the technic of this operation.

In all serious cases of inflammation of the carpal flexors, whether tenotomy has been performed or not, the subject needs a long period of rest subsequent to treatment. In fact, three or four months at pasture is necessary to permit of recovery and this where no congenital deformity has predisposed the subject to such affection of the flexors. Return to work must be gradual and the character of the work such as to enable the animal to become inured to service without a recurrence of the trouble if possible.

It follows then, that tenotomy, here as in other cases, is not practical from an economic viewpoint, unless the animal be of sufficient value to justify the long period of rest for recovery. Tenotomy is not of practical benefit unless ample time is allowed for regeneration of divided tendinous tissue.

Fracture and Luxation of the Carpal Bones.

Etiology and Occurrence.—Fracture of the carpal bones is of infrequent occurrence in horses and, when it does occur, it is usually due to injuries, and because of their nature (resulting as they generally do from heavy falls or in being run over by street cars or wagons), a comminuted fracture of one or more bones exists. The accessory carpal bone (trapezium) is said to be fractured at times without being subjected to blows or like injuries, but this is exceptional.

Luxations of the carpal joint are of rare occurrence, and very few cases of this kind are on record. Walters[18] reports a case of carpometacarpal luxation in a pony wherein reduction was spontaneous and an uneventful recovery followed. His reason for reporting the case, as he states, is its rarity.

Symptomatology.—Fractures of the carpal bones as they usually take place are diagnosed without difficulty. Because of their usually being comminuted, abnormal movement of the joint is possible. Such movement is not restricted and flexion of the leg at the carpus in any direction is possible. Crepitation is readily detected and frequently these fractures are of the compound-comminuted variety.

In fracture of the accessory carpal bone (trapezium) or in fracture of any other single bone when such exists, there is no increase in the movement of the joint. The accessory carpal bone may be readily manipulated and when fractured, its parts are more or less displaced. Recognition of fracture of any other single carpal bone must be done by detecting crepitation unless it be a compound fracture, whereupon probing is of aid in establishing a diagnosis.

Carpal luxation when present is to be recognized by finding the apposing carpal bones joined in an abnormal manner—that is, out of position. There is restricted or suspended function of the joint, and in the cases recorded, no difficulty has been experienced in making a diagnosis. The carpometacarpal portion of the articulation is the part which is usually affected.

Prognosis and Treatment.—There is no chance for complete recovery in the usual case of carpal fracture because of the fact that there results sufficient arthritis to destroy articular cartilage beyond repair. In the average instance, because of arthritis which persists for a considerable length of time, more or less ankylosis results. At best, one can only hope for partial recovery, that is to say, the member may regain its usefulness as a weight-supporting part, but because of restricted or abolished joint function, locomotion is more or less difficult. Exostoses, articular and periarticular, occur and the carpus usually becomes a large immobile articulation. There is danger of infection resulting in simple carpal fractures and, needless to say, in a compound-comminuted fracture of the carpus, infection usually occurs and a fatal outcome is probable.

When treatment is instituted, antiseptic precautions are taken in handling the compound fractures, and in any case immobilization of the parts is sought. Here, as has been previously pointed out, it is best to employ leather splints, so that a maximum degree of rigidity with a minimum of distress and inconvenience to the patient will result. The leg must be bandaged from the hoof upward, making use of a sufficient amount of cotton to ensure against pressure-necrosis. The leather splints are placed mesially and laterally and, of course, need to extend as high as the proximal end of the radius. Subjects must be kept in slings until union of bones has become established, and as a rule there will then exist marked ankylosis.

There is no particular difference in the handling of carpal luxation and dislocation of other bones. Where ligaments have not been destroyed to the extent that reduction is of no practical use, the parts are kept immobilized, if thought necessary. Later, vesication of the whole pericarpal region is done and the subject allowed exercise at will.

Carpitis.

Etiology and Occurrence. — Inflammation of the carpus is caused by contusions, such as are occasioned in falling, by kicks by striking the carpus against objects in jumping and sometimes by striking it against the manger in pawing. The condition is of rather frequent occurrence.

Symptomatology. — Evident symptoms of inflammation in carpitis are always present — hyperthermia, supersensitiveness and swelling. Also, there exists lameness which is characterized by an apparent inability to flex the leg, and there is circumduction of the leg as it is advanced because in this way little if any flexion of the carpus (which increases pain) is necessary.

Depending upon the nature of the cause, there occurs a marked difference in the character and amount of swelling.

Fig. 11 — Pericarpal inflammation and enlargement due to injury.

Naturally, when much extravasation of serum and blood takes place, there is occasioned a fluctuating swelling which is usually less painful to the subject upon manipulation than is a dense inflammatory change without marked extravasation.

In acute carpitis, there is present, then, a very painful condition which involves the articulation, causing marked lameness, disturbance of appetite and some elevation of temperature.

Chronic cases do not occasion serious pain or constitutional disturbances, but do interfere with locomotion in direct proportion to the existing articular inflammation and periarticular hypertrophy of ligamentous and tendinous structures.

Treatment.—If possible, keep the subject absolutely quiet, employing the sling if necessary. During the first stages of inflammation, the application of ice packs to the affected parts, is of marked benefit. At the end of forty-eight hours, hot applications may be used and this treatment continued throughout several days. Anodyne liniments are of service and should be employed throughout the acute stage of inflammation during intervals between the hydrotherapeutic treatments.

As inflammation subsides, a counterirritating application such as a suitable liniment and later blistering or line-firing is helpful in stimulating resolution.

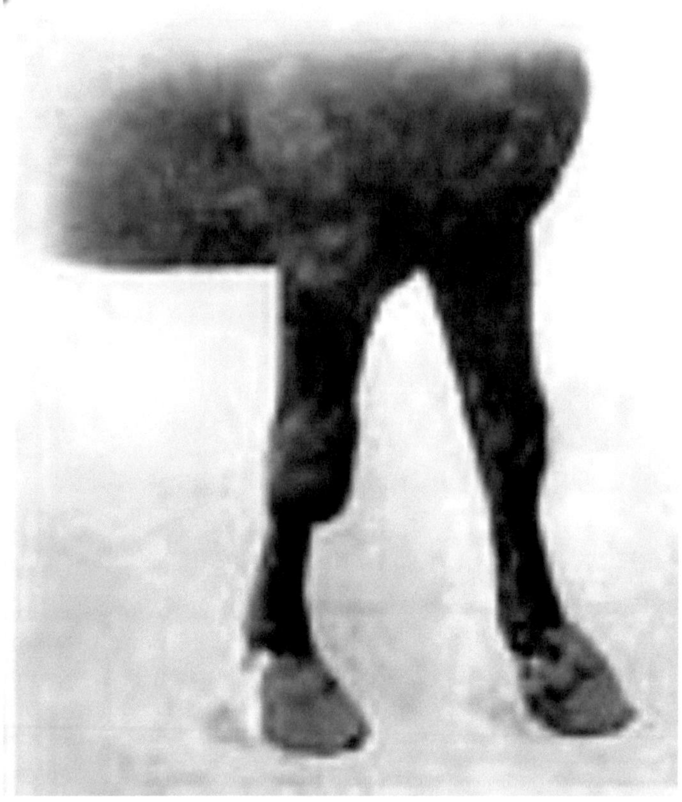

Fig. 12—Hygromatous condition of the right carpus, also distension of sheaths of extensor tendons of both fore legs.

Open Carpal Joint.

Anatomy.—The carpal bones as they articulate with one another and with the radius and metacarpal bones, as classed by anatomists, form three distinct articular parts of the joint as a whole and are known as radiocarpal, intercarpal and carpometacarpal. These three pairs of articulating surfaces are all enclosed within one capsular ligament. On the anterior face of the bones, the capsular ligament is attached to the carpal bones in such manner that an imperfect partitioning of the three joint compartments is formed. Posteriorly, the capsule is very heavy and forms a sort of padding over the irregular surfaces of the bones, and also its reflexions constitute the sheaths of the flexor tendons. The anterior portion of the capsular ligament forms sheaths for the extensor tendons, and both portions of the joint have an attachment around the distal end of the radius and another at the proximal end of the metacarpal bones.

Fig. 13—Carpal exostosis in aged horse.

Etiology and Occurrence.—Puncture wounds of any kind may serve to perforate the joint capsule and such traumatisms are occasioned by falls, kicks and in various ways in runaway accidents, and open carpal joint may follow.

Symptomatology.—The pathognomonic symptoms of the existence of an open joint is the exposure to view of articular surfaces of bones or noting the escape of synovia from the joint capsule. As has been previously referred to, there always exists a peculiar suspension of carpal flexion in all cases of carpitis.

Non-infective wounds which may cause open joint are not necessarily productive of an active carpitis—a synovitis may be the extent of the disturbance. Unlike synovitis, which may characterize a non-infectious penetrative wound of the capsular ligament, septic arthritis which may supervene is a very painful inflammatory disturbance. It is characterized by all of the symptoms which attend the case of open joint and synovitis plus the obvious manifestation of great pain. There is an elevation of temperature of from two to five degrees above normal; circulation is accelerated; the pulse is bounding; respiration is hurried; there is an expression of pain as indicated by the physiognomy; and because of rapid erosive changes of cartilages which take place, there is soon so much of the articulation destroyed that death is inevitable. Death is usually due to generalization of the arthritic infection.

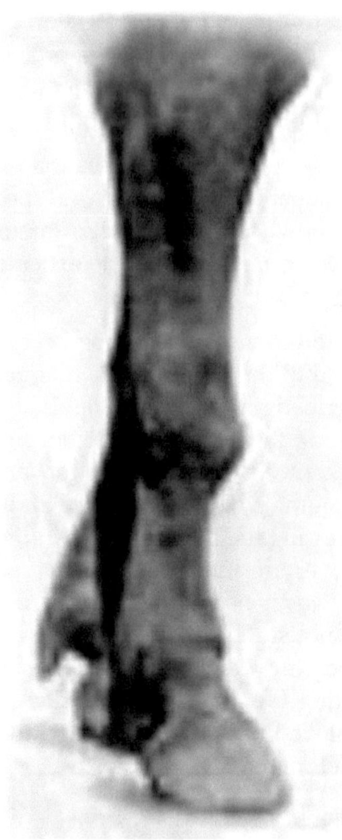

Fig. 14—Exostosis of carpus
resultant from carpitis.

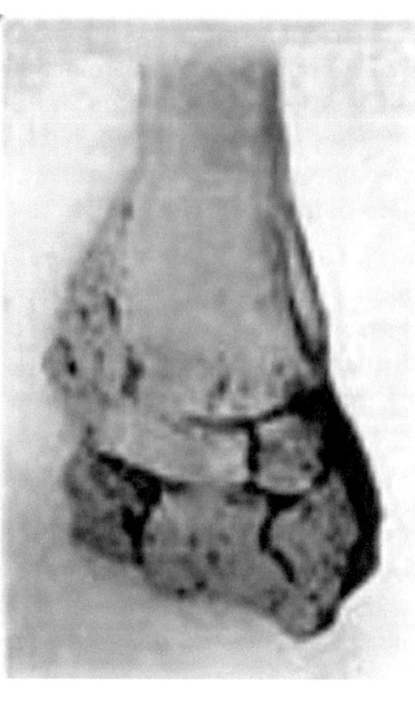

Fig. 15—Distal end of radius.
Illustrating the effects of
chronic carpitis.

In the meanwhile, if the character of the infectious material is not so virulent, the disease will take on a slower course and the subject may experience laminitis from supporting weight upon the sound member, or because of continued recumbency, decubital gangrene and emaciation sometimes cause death. If the subject does not soon succumb, it is compelled to undergo days or even weeks of unnecessary suffering, and too often in such cases, it is later deemed advisable to destroy the animal because of the cost of continuing treatment until the horse is serviceable. Therefore, it is evident that

149

when such joints as the carpus or tarsus are open and infection exists, if they are not promptly treated and the infectious process checked, it is neither humane nor practical to prolong treatment.

Distinction must be made between the different joints when infected as the condition is much more serious in some cases than in others. All things considered, perhaps open joints rank, with respect to being serious cases as follows: elbow, navicular, stifle, tarsus, carpus, fetlock and pastern. This, of course, is restricted to articulations of the locomotory apparatus.

Treatment.—Preliminary care in the treatment of an open carpal joint, is the same as has been described in this condition as it affects the scapulohumeral articulation described on page 65. Likewise the further treatment of such cases is along the same lines except that where it is possible, the parts are kept covered with cotton and bandages. However, in some cases, animals have been successfully treated without bandaging and by keeping the patient in a standing position and on pillar reins until recovery resulted. Such cases were of the non-infectious type and recovery was possible within three or four weeks. Further, the condition is not sufficiently painful in such instances as to prevent the subjects bearing weight with the affected member; hence, no danger of resulting laminitis is incurred. And finally, where bandages are not employed, the frequent use of antiseptic dusting powders is substituted for cotton as a protector.

When bandaged, such wounds need dressing more or less frequently, as individual instances demand. The purulent infective inflammation of a carpal joint will require daily dressing; whereas, in other cases (non-infective), semi-weekly change of bandages is sufficient. Equal parts of boric acid and exsiccated alum constitute a suitable combination for the treatment of these cases, and this powder should be liberally employed. Tincture of iodin may be injected into the joint capsule where there is provision for its ready evacuation, as conditions seem to require. Daily injections for three, four or five days, are not harmful and will control infection in many instances.

Thecitis and Bursitis.

Etiology and Occurrence.—The thecae and bursae of the leg are several in number. In the carpal region, the flexors of the phalanges are contained together in the carpal sheath, and this is the principal theca in the carpal region. Each of the tendons is provided with synovial sheaths which are subject to inflammation and occasionally synovitis and distension of these synovial sheaths occur.

Because of faulty conformation, some animals are subject to inflammation of these sheaths, and all forms of strenuous work which taxes the tendons greatly is apt to result in synovitis. Direct injury such as blows may be the cause of synovial distension of thecae and the affection is to be seen in all horses that have done much fast work on hard road surfaces or pavements.

The usual case as it occurs in practice is a non-infective synovitis, but where puncture wounds cause the trouble, an infectious inflammation obtains.

Symptomatology.—No trouble is experienced in diagnosing distension of tendon sheaths, for the affection is very palpable. During acute inflammatory stages of this affection, some lameness is present—in infectious inflammation lameness is intense. Local heat and pain upon manipulation are readily discernible in all acute cases. And finally, where there is reason for doubt, an aseptic exploratory puncture of the wall of the fluctuating enlargement may be made with a suitable trocar, and the discharging synovia will be proof of the existence of synovial distension.

After the affection becomes subacute or chronic, no lameness or inconvenience is manifested, and the condition is undesirable only because of its being a blemish.

Treatment.—Acute non-infectious synovial distension of tendon sheaths is treated by aspirating as much synovia as possible from the affected theca (this is, of course, done under strict asepsis) and by means of bandages, a uniform degree of pressure is kept over the parts for ten days or two weeks. The patient is kept quiet and in the course of two weeks an active blistering agent is employed over the region affected. Usually, at the end of a month's time, complete recovery has taken place and the subject may be gradually returned to work.

When synovial distensions are of long standing, it is necessary to take special precautions to check excessive secretion of synovial fluid, and, also because of the atonic condition of the tissues affected, resolution is tardy. In addition to aspirating synovia, the introduction of equal parts of alcohol and tincture of iodin into the theca is necessary. The quantity of this combination injected, depends upon the size of the sheath affected and the amount of synovia retained at the time injection is made. Experience is necessary to judge as to this part of the work, but one may consider that a quantity between three and ten cubic centimeters of equal parts of tincture of iodin and alcohol constitutes the proper amount to employ. Where much synovia is contained within the sheath at the time of injection, there occurs great dilution of the agent injected and consequently less irritation results.

The object of such injections is to check synovial secretion, and this is sought by the local effect of iodin in contact with the secreting cells together with the reactionary swelling which occasions pressure. An increase in the local blood supply also follows. In all cases where it is possible to employ suitable bandages, this should be done. The ordinary derby bandages serve well and if their use is continued for a sufficient length of time, good results follow.

There are other methods of treating these affections, and each has its advantages and disadvantages. Line-firing, instead of the vesicant is made use of by some, but the object desired is the same and results obtained are similar.

Sheaths may be opened surgically by means of a knife, and the removal of a portion of the wall of distended and atonic tendon sheaths is possible. These operations belong to the realm of surgery and are not properly a part of this treatise. However, in passing, it may be said that if a perfect technic is possible in doing the last named operation, a permanent recovery is the outcome.

Fracture of the Metacarpus.

Etiology and Occurrence.—As the result of all sorts of violence, such as falls and injuries in accidents of various kinds wherein the metacarpals are subjected to contusions, fractures may result. In the horse it is unusual for fracture of one of the small metacarpal bones to take place without there being at the same time a fracture of the third (large) metacarpal bone.

Classification.—Fractures of the metacarpal bones as they occur, are as likely to be compound as simple, and the multiple and comminuted varieties are occasionally observed. The manner in which the third (large) metacarpus is fractured, largely determines the outcome in any given case.

Symptomatology.—Abnormal mobility of the broken parts of bone and crepitation mark fracture of the metacarpus, and the condition is easily diagnosed. In many instances, when compound fracture exists, broken ends of bone are protruding through the skin. No weight is borne upon the fractured member ordinarily, although during the excitement occasioned by runaways, horses are sometimes seen to support weight with a broken leg even when the protruding bone is sunk into the ground in so doing.

Prognosis.—Generally speaking, fractures other than the simple-transverse in young animals, are considered unfavorable cases. With the metacarpus, however, there are instances where compound fracture occurs in colts that justify treatment. But in all cases of compound fracture, the element of infection in addition to the increased difficulty in maintaining immobility of the broken bone, creates almost insuperable difficulties in the average instance. And unless the practitioner distinctly explains to his client the various reasons which make treatment an economic impracticability, dissatisfaction is likely to follow if treatment is instituted without such an understanding.

Treatment.—Perfect apposition of the broken ends of bone is easily effected and less difficulty is encountered in maintaining such relations in metacarpal fractures than in fractures of the radius. However, reduction and immobilization of this as in all fractures, must be done without delay. In simple fracture, the metacarpus is covered with enough cotton to pad the parts, and this is retained in position by bandages. Splints of heavy leather or of thin pieces of tough flexible wood are placed on each side of the leg and firmly held in position with bandages. Bandages may be put on in layers and a coating of glue applied over each layer if this is thought necessary. The advantage gained in using glue or other adhesive materials is that the cast thus formed is more rigid than where such material is not employed. On the other hand, all elasticity is lost as soon as the cast adapts itself to the contour of the extremity, and because of this rigidity, it is doubtful if anything is gained by the incorporation of glue, except in the way of added strength of the cast. Since the animal does not walk upon the broken leg, it is possible to employ splints of suitable materials which are retained in position without glue and frequent readjustment of a part of the immobilizing apparatus is possible. This is impossible with casts.

In compound fractures, provision ought to be made for dressing the wound of the soft structures. This entails adjusting the splints in such manner that one splint may be retained and others removed for dressing the wound and readjusted as often as wound dressing is necessary.

Splints.

By this term is meant a condition where there exists an exostosis which involves usually the second (inner small) and third (large) metacarpal bones. While an exostosis involving any one of the splint bones, even when directly caused by an injury, is called a "splint," the term is employed here, in reference to exostoses not due to direct injury such as in contusions.

Etiology and Occurrence. — This condition is one wherein there is osseous formation following a periostitis and the region of the upper portion of the second (inner small) metacarpal bone is the usual site of the exostosis. There is incited an inflammation of the periosteum at the site of the interosseous ligament which attaches the small to the large metacarpal bone. This ligament is involved in the inflammatory process, and according to Havemann, whose view is supported by Moller, this inflammation is the origin of the trouble.

Various theories attempting an explanation of the frequent affection of this one certain part so regularly involved have been offered, but no proof of the correctness of any exists. It follows, however, that splints occur in young animals; that the affection seldom starts in subjects that are ten years of age or older, and that when the exostosis has formed, lameness usually subsides. Anything which will cause undue strain or irritation of the metacarpal bones in young animals, is quite apt to result in a splint being formed. Concussion such as is caused by fast work on hard roads, or work on rough or irregular road surfaces which cause unequal distribution of weight, will cause splint lameness and exostosis follows.

Fig. 16—Posterior view of radius (right) illustrative of effects of splint. Note the extent of exostosis.

Course.—Because of the peculiar manner in which the second and third metacarpal bones articulate in young animals, until the bones become ossified and permanently joined, the inflammation which attends the acute stage of this affection, causes lameness. Later, unless an unusually large exostosis is formed, which may cause a constant irritation due to its size and juxtaposition to the carpus, lameness is discontinued.

Symptomatology.—Lameness is usually the first manifestation of this disorder, and the thing which characterizes splint lameness is its peculiar intermittence. There is a mixed form of lameness which may not be in evidence when an affected animal is started on a drive, but which is marked after the subject has gone some distance. The animal may, however, go lame throughout the whole of a drive and continue to be lame for several days or weeks in some cases. It is noticeable that lameness is augmented or produced when the subject travels on rough road surfaces and that little or no difficulty is encountered when roads are smooth.

The heavy brachial fascia is inserted in part to the head of the second metacarpal (inner small) bone together with the oblique digital extensor (extensor metacarpi obliquus) and this explains the reason for pain being manifested during extension of the member.

Before there is a visible exostosis, supersensitiveness is readily recognized upon palpation of the parts, if careful comparison is made between the sound and unsound members. However, frequently splints occur on both forelegs at the same time and in some instances exostoses are several in number upon each member affected. In some instances, the affection involves the outer splint bone and no evident involvement of the inner one exists.

Treatment. — At the onset complete rest should be provided and the local application of some good cataplasm is in order. A stimulating liniment is beneficial when employed several times daily and massage is also quite helpful. Later, the application of a blistering ointment is good treatment. The use of the actual cautery stimulates prompt resolution, but there is seldom any resorption of products of inflammation following firing. Whereas, in cases where other treatment is begun early, there usually follows considerable diminution in the size of the exostosis. A rest of four or five weeks is necessary and very young animals should not be put to work too soon, if the character of the work is such as to induce a recurrence of the trouble.

Many cases are treated successfully in draft types of animals (where the subjects are not kept at work that occasions serious irritation to the affected parts) by blistering the exostosis repeatedly and allowing the animals to continue in service. In such cases, it is unreasonable to expect to check the size of the exostosis and, of course, such methods are not employed where lameness causes distress to the subject.

Firing usually causes prompt recovery from lameness and is a dependable manner of treating such cases but there remains more blemish following cauterization than where vesication is done.

OPEN FETLOCK JOINT.

This condition, because of the frequency with which it occurs may be taken as typal, from the standpoint of treatment and results obtained therefrom. While it serves to constitute a basis from which other joints, when open, are to be considered, due allowance must be made for the fact that, as has been previously mentioned, some articulations when open constitute cause for grave consequences; while with others an open capsule, even when infected, does not cause disturbance enough to be classed as difficult to handle. Moreover, the fetlock joint is admirably suited, anatomically, to bandaging; and when wounded, is easily kept protected by means of surgical dressings. This fact is of great importance in influencing the course and termination in any given case of open fetlock joint and should not be forgotten.

There is no logical reason for comparing the pedal joint with the pastern on the basis that it may also be completely and securely bandaged. Open navicular joint does not occur, as a rule, except by way of the solar surface of the foot, and the introduction of active and virulent contagium is certain to happen; consequently, an acute synovitis quickly resulting in an intensely septic and progressively destructive arthritis soon follows in perforation of the capsule of the distal interphalangeal articulation.

Etiology and Occurrence. — Wounds of the fetlock region resulting in perforation or destruction of a part of the capsular ligament are caused by all sorts of accidents, such as wire cuts, incised wounds occasioned by plowshares, disc harrows, stalk cutters and other farming implements. In runaways the joint capsule is sometimes punctured by sharp pieces of wood or other objects. In horses driven on unpaved country roads the fetlock is occasionally wounded by being struck against the sharp end of some object, the other end of which is firmly embedded in the ground. In one instance the author treated a case wherein the fetlock joint was perforated by the sickle-guard of a self-binder. In this case there occurred complete perforation causing two openings through the *cul-de-sac* of the joint. Such wounds are produced by implements which are, to say the least, non-sterile, and this perforation of the uncleansed skin conveys infectious material into the joint capsule. Yet in many instances, especially in country practice, no infectious arthritis results where cases are promptly cared for.

Symptomatology.—A difference in the character of symptoms is evidenced when dissimilar causes exist. Small penetrant wounds which infect the synovial membranes cause infectious arthritis in some cases, whereas a wound of sufficient size to produce evacuation of all synovia will, in many instances, cause no serious distress to the subject, even when not treated for several days. If it is not evident that an open joint exists and the articular cavity is not exposed to view a positive diagnosis may be early established by carefully probing the wound. In some cases where a small wound has perforated the joint capsule, swelling and slight change of relation of the overlying tissues may preclude all successful exploratory probing. In such instances it is necessary to await development of symptoms. Twenty-four hours after injury has been inflicted, there is noticeable discharge of synovia which coagulates about the margin of the orifice, where synovial discharge is possible. Particularly evident is this accumulation of coagulated synovia where wounds have been bandaged—there is no mistaking the characteristic straw-colored coagulum which, in such cases, is somewhat tenacious.

No difference exists between other symptoms in infectious arthritis caused by punctures, and non-infectious arthritis, excepting the intensity of the pain occasioned, the rise in temperature, circulatory disturbances, etc.; all of which have been previously mentioned.

Treatment.—Just as has been stated in discussions on the subject of open joint, probing or other instrumentation is to be avoided until the exterior of the wound and a liberal area surrounding has been thoroughly cleansed—too much importance can not be placed on this preliminary measure. In cases of open joint where ragged wound margins exist and the interior of the joint capsule is contaminated, much time is required to thoroughly cleanse all soiled parts. In some instances an hour's time is required for this cleansing process after the subject has been restrained and prepared. In order to thoroughly cleanse these delicate structures without doing them serious injury, one ought to be skillful and careful in all manipulations of the exposed parts of the joint capsule.

The general plan of treatment, after preliminary cleansing has been accomplished, has been outlined on page 66 in the consideration of scapulohumeral joint affections. The injection of undiluted tincture of iodin in ounce quantities, it must be remembered, is not to be done unless there is provision for its free exit. Where good drainage from the joint cavity exists all infected wounds should be thus treated, and this treatment may be repeated as conditions seem to require — until infection is checked.

If daily injections are necessary, dilution of the tincture of iodin with an equal amount of alcohol is advisable in order to avoid doing irreparable damage to the articular cartilages and synovial membranes.

An antiseptic powder composed of equal parts of boric acid and exsiccated alum is employed to protect the wound surfaces and the margins, and the parts are then bandaged. In bandaging wounds of this kind a liberal amount of cotton should be employed, and after a large surface surrounding the wound has been thoroughly cleansed, it must be so kept thereafter. This is impossible, if one uses a small amount of cotton, particularly if such meager quantity of dressing material is carelessly wrapped in position with an insufficient amount of bandage material. Mention, without description of the elemental problem of applying cotton and bandages to a wound, would be sufficient, were it not that this is a very important part of the handling of such cases, and many practitioners are not only thoughtless in this part of their work, but also apparently careless. What does it profit to prepare a part and cleanse a wound with painstaking care and then neglect to take every possible precaution to prevent its subsequent contamination?

In the handling of open joint capsules where the perforation of the capsular ligament is small and discharge of synovia does not immediately follow, there is presented a problem which is difficult to decide upon and that is the manner in which such wounds are to be handled. One hesitates to enlarge such openings to drain or irrigate the capsule when there is no proof that serious trouble will follow because of infectious material which has probably been introduced at the time the wound was inflicted. It is especially difficult to decide upon the manner of handling such cases where the tarsal joint is wounded, although one hesitates to invade any joint to the extent of incising its capsule, unless there is urgent need of so doing.

Frost[19] offers the following suggestion in such instances:

The treatment recommended by us for open joints, in which we wish to prevent ankylosis, is, first, to shave all hair from the area surrounding the wound, following with a thorough cleansing of the skin and disinfection of the wound, and then to inject a twenty per cent Lugol's solution in glycerin into the wound. This should be repeated two or three times a day, each time enough of the solution being injected to fill the joint capsule, thereby securing the flushing effect. As this solution does not cause irritation to the tissue and yet is a strong antiseptic, it serves to shorten the period of congestion and inflammation and to overcome the infection without causing a destruction of the secreting membrane until the external wound has had time to heal. The injection of this solution seems to retard the excessive secretion of synovia. The larger the joint capsule and the smaller the external wound, the longer our antiseptic will remain in contact with the inflamed tissues as the glycerin, being thick, does not flow through a small opening.

After-care.—Following the initial cleansing and treatment of open joint, subsequent dressing is necessary as frequently as conditions demand. If the parts are badly infected and profuse discharge of pus exists a daily change of dressings is necessary. In the average instance, however, semi-weekly treatments are sufficient. And in many instances where one is obliged to travel a considerable distance to handle the affected animal one weekly dressing of the wound will suffice after the second treatment.

The same general plan of treatment concerning the subject's comfort that has been previously mentioned in arthritis, is carried out here. A further and detailed consideration of the subject of handling of open joints follows.[20]

* * * Such wounds may be classified in two general groups as follows: First, wounds in which the trauma has exposed the articulation to view, and second, those the result of punctures, in which the external wound is small and free drainage is lacking.

Wounds in which the articulation is exposed to view have drainage either all ready provided for, or it is established without hesitancy surgically. With free drainage thus established there is little or no chance for the adjacent tissues to become infiltrated with infected wound discharge. This prevents an extension of the injury and the establishment of a good field for the growth of anaerobic bacteria.

Open joints caused by punctures, unless the puncture is aseptic, produce a swelling which is more painful than is the open wound which exposes the joint to view. Especially is this true if the puncture is of small diameter, allowing the tissues to partially close the opening immediately after the wound has been made. Where drainage is lacking there follows an exudation which congests the tissues surrounding the injury and all factors favoring germ growth are present. It is perhaps advisable to establish good drainage in such cases as soon as a diagnosis is made.

It is not always an easy matter to recognize an open-joint, when first made, but twelve to twenty-four hours later there is no cause for doubt. The condition is then a very painful one; lameness is excessive; there is rise in temperature; acceleration of the pulse and manipulation or palpation of the region affected, occasions great pain.

The treatment of open joints must be varied to suit the disposition of the animal, the nature and location of the injury, the length of time intervening between the infliction of the wound and the first attention given, and the surroundings in which the patient is kept.

In each and every case in which there exists an open wound the surface surrounding the wound is cleansed thoroughly, the hair is shaved if possible, and the margin of the wound is curretted and cleansed thoroughly with antiseptic solutions.

If there is evidence that the articulation contains infective material, it is washed out with copious quantities of peroxide of hydrogen — usually as much as six or eight ounces. This is followed by injection of an ounce or two of tincture of iodin. Even though the joint appears to be clean some tincture of iodin is used, as it checks the secretion of synovia and is, in every way, beneficial. Care is taken to apply the iodin also to the surface immediately surrounding the wound. The entire wound is then covered with a dusting powder composed of zinc oxide, boric acid, exsiccated alum, phenol and camphor.

This powder is used in abundance and the wound is then covered with a heavy layer of absorbent cotton and well bandaged. This bandage is not disturbed for at least three days and may be left in place for a week. In cases in which it is necessary to keep the dressing on for a week, or in cases where the patient is, through necessity, kept in quarters that are wet or unclean, the first bandage is covered with a layer of oakum which has been saturated in oil of tar and this in turn is held in place by means of several layers of bandages. The bandages are also saturated with oil of tar.

In from one to two months wounds so treated, unless they are foot-wounds, will be ready to dress without being bandaged. It is ordinarily unnecessary to dress foot-wounds oftener than every second week after the discharge of synovia has ceased. When the wound has filled with granulation, a protective dressing is applied which is rendered water proof by the use of bandages covered with oil of tar. The patient can now be turned out for a month or six weeks without disturbing the dressing. After the removal of the bandages, the only treatment necessary is an occasional application of some mildly antiseptic ointment.

Except in nail pricks of the foot, occasioned by punctures, a five per cent tincture of iodin is injected into open joints, if the wound remains sufficiently open, and this treatment is continued so long as there is a discharge of synovia. Surgical drainage is established if it is considered practicable and the remainder of the treatment is about the same as for wounds which are open.

Open joints occur in horses at pasture and are sometimes not discovered until several days or a week after the injury, and in some instances the wounds are filled with maggots. The only difference in the treatment of these cases is that more time and care is taken in cleansing the wound, more curetting is necessary, and after cleansing the wound with peroxide of hydrogen, the joint is thoroughly washed out with equal parts of tincture of iodin and chloroform. This is followed by the injection of a quantity of seventy-five percent alcohol and the wound is dressed and bandaged as already described. At each subsequent dressing of infected wounds so treated less suppuration is noticed and the synovial discharge usually ceases in from one to two months.

About *ninety percent of all cases of open joint make complete recoveries,* about four per cent partially recover and six per cent are fatal. Among the fatal cases are the open joints with complications as severed tendons, those occasioned by calk wounds in horses that are stabled, and nail punctures of the feet. The following report of twelve favorable cases is taken from a record of sixty-two cases. The favorable ones are reported, chiefly because there are now enough reports on record of such cases which have terminated fatally.

Case 1.—A gray gelding used as a saddle pony received a horizontal wire cut laying completely bare the scapulohumeral articulation. The margins of the wound were cleansed as heretofore described, a drainage was provided surgically, tincture of iodin was injected and the wound was covered with equal parts of boric acid and exsiccated alum. The horse was kept tied and a diluted tincture of iodin was injected into the wound once daily and the powder applied often enough to keep the wound covered. The case made a complete recovery and the pony was again in service within sixty days.

Case 2.—A twelve-hundred-pound bay mare with an open carpal joint. The wound was an open one about two and one-half inches in length, and made transversely and when the member was flexed the articular surface of the carpal bones were presented to view. An ounce of tincture of iodin was injected into this joint after having cleansed the margin of the wound and the mare was cross-tied in a single stall to keep her from lying down. The owner was instructed to keep the outside of the wound powdered with air slaked lime and a very unfavorable prognosis was given.

I heard nothing further from this case until fifty-nine days from the date of the injury, when I met the owner driving this mare to a buggy. The wound had healed by first intention and at that time so little cicatrix remained that it was difficult to find it.

Case 3.—A brown mare with an open fetlock joint due to a spike-nail puncture. Lameness was excessive, and joint greatly swollen. Tincture of iodin was injected into the wound and towels dipped in hot antiseptic solutions were applied for several hours daily until the acute stage had passed. Later the mare was turned out to pasture and a vesicant was applied once or twice a month until recovery was complete which was in about six months.

Case 4.—A four-year-old bay mare having a wire-cut which opened the tarsus joint was treated as heretofore described. The wound was kept bandaged for about two weeks and later it was dressed without being bandaged. In ninety days she had completely recovered.

Case 5.—A twelve-year-old mare with an open fetlock joint due to a puncture wound. The margins of the wound were cleansed and the external wound enlarged to facilitate drainage. Tincture of iodin was injected; the wound was bandaged and dressed for a month in the manner heretofore described, when all discharge had stopped. A vesicant was applied; the mare was put to pasture and within sixty days from the date of the injury she was being driven on short trips.

Case 6.—A two-year-old brown gelding with a wire-cut on the left front foot. The wound extended down through the sole and opened the navicular joint. This colt was very wild and it was necessary to tie it down each time the wound was dressed. The wound was dressed weekly for a month and less frequently thereafter. It was handled eight times; the last dressing was left in place until worn out. Six months later the colt was practically well, a very little lameness being shown when walking on frozen ground.

Case 7.—A seven-year-old saddle-horse weighing eleven hundred and fifty pounds received a wound of the tarsus, laying bare the articular surfaces of a part of the joint. It was impossible to keep this wound bandaged because of the restless disposition of the subject. Injections of a dilute tincture of iodin were employed every second or third day for a month and the wound was kept covered with the antiseptic dusting powder referred to heretofore. In five months complete recovery had taken place, with the exception of a stubborn skin disturbance which was successfully treated six months after the wound was inflicted. The horse is still in use and is absolutely free from lameness.

Case 8.—A two-year-old brown gelding with a wire-wound opening the scapulohumeral joint. This wound was large enough to expose to view the articular portion of the humerus. The same treatment as that given case No. one was instituted and in ninety days the colt was practically well.

Case 9.—A three-year-old bay filly was found at pasture with one fore foot badly injured. The owner intended to destroy her, but a neighbor prevailed upon him to have her treated. Apparently the wound was of about a week's standing and in a very bad condition, filled with maggots and dirt. Both the navicular and coronary articulations were open. This wound was cleansed in the usual manner and the owner cared for the case the balance of the time because the distance from my office was too great to give her personal attention. She made an almost complete recovery in five months.

Case 10.—At two-year-old mule with an open navicular joint due to a barbed wire wound. Usual care was given this case and in five months recovery was complete and little scar is to be seen. This case received seven treatments.

Case 11.—An eighteen-months-old colt at pasture was found down and unable to rise without help. In addition to several wounds of lesser importance there was a large wound on the inner side of the elbow, the joint was open and the entire leg was greatly swollen and in a state of acute infectious inflammation. The colt could not walk, its temperature was 105°, pulse was rapid and respiration was a little hurried. After advising the owner to put the poor animal out of its misery I left the place. Four days later the owner came to my office and asked if he could borrow some old shears to "trim off some loose hide from that colt." He left the colt in the pasture and all the care it received was the regular application of a proprietary dusting powder. It made a complete recovery.

Case 12.—A family mare, heavy in foal, received a vertical wound of the fetlock joint inflicted by a disc-harrow. The *cul-de-sac* of the ligament of this joint was opened freely. The wound was dressed in the usual manner and again three days later when no suppuration had taken place. Four days later the patient gave birth to a colt and suckled it right along through her convalescence. This wound healed by first intention and seventy-nine days from the date of the injury the mare was driven to town, two and one-half miles distant, and showed but little lameness.

Phalangeal Exostosis (Ringbone)

This term is applied to exostoses involving the first and second phalanges (suffraginis and corona), regardless of their size, extent or location. It is a misnomer, in a sense, and the veterinarian is frequently obliged to spend considerable time with his clients in order to convince them that a spherodial exostosis of the proximal phalanx, in certain cases, is in reality "ringbone," even though there exists no exostosis which completely encircles the affected bone.

Etiology and Occurrence.—Exostosis of the first and second phalanges is usually due to some form of injury, whether it be a contusion, a lacerated wound which damages the periosteum, or periostititis and osteitis incited by concussions of locomotion, or ligamentous strain. Practically the only exception is in the rachitic form of ringbone which affects young animals.

There are predisposing causes that merit consideration, chief among which is the normal conformation of the coronet joint. This proclivity is constant; the normal interphalangeal articulation is an incomplete ginglymoid joint and while its dorso-volar diameter is great, this in no wise compensates for its disproportionately narrow transverse diameter. The pivotal strain which is sometimes thrown upon this articulation when an animal turns on one foot, as well as the tension which is put on the collateral ligaments when the inner or the outer quarter of the foot rests in a depression of the road surface, tends to detach the insertion of these ligaments or to cause fibrillary fractures of their substance.

Short, upright, pasterns receive greater concussion during fast travel on hard roads than do the longer more sloping and well formed extremities. Those who are advocates of the theory that this type of osteitis with its complications has its origin in the articular portion of the joint, claim that the upright pastern constitutes an important tendency toward ringbone. Howbeit, ringbone is an active, serious and frequent cause of lameness and it affects animals of all ages and occurs under various conditions. Horses having good conformation and kept at work wherein no great amount of strain is put upon these parts, are occasionally victims of this affection.

Classification.—The arrangement employed by Moller[21] is intensely practical and logical. He considers ringbone as *articular, periarticular, rachitic* and *traumatic*. A mode of classification that is common and in a practical way, good, is, high and low ringbone. When prognosis is considered, for instance, it is very convenient to state that the chances for recovery are much better in high ringbone than in low ringbone. The classification of Möller will be followed here.

Fig. 17—Phalangeal exostoses.

Symptomatology.—In all forms of incipient ringbone except rachitic, the first manifestation of its existence, or of injury to the ligaments in the region of the pastern joint which causes periostitis, or affections of the articular portions of the proximal inter-phalangeal joint, is lameness. Lameness which typifies ringbone is of the supporting-leg variety and by compelling the subject to step from side to side, marked flinching is observed, especially in periarticular ringbone; causing the affected animal to turn abruptly on the diseased member, using it as a pivot, likewise accentuates the manifestation. In fact, many subjects that exhibit no evidence of locomotory impediment while walking or trotting in a straight line on a smooth road surface, will manifest the characteristic form of lameness from

170

ringbone when the aforementioned side to side movement is performed.

When the manner in which pain is occasioned is considered, it
will be understood why lameness is intermittent in the early stages
of this affection and may even be unnoticed by the driver. An animal may travel on a smooth road without giving evidence of any
inconvenience, but as soon as a rough and irregular pavement or
road surface is reached, will limp. As the subject is driven farther on
level streets the lameness may disappear. This intermittent type of
lameness may continue until there is developed a large exostosis, or
until articular involvement causes so much distress during locomotion that lameness is constant. On the other hand, resolution may
occur during the stage of periosteal inflammation, or, an exostosis
forms which causes no interference with function.

Fig. 18—Rarefying osteitis in chronic ringbone and ossification of
lateral cartilages.

Before there is evidence of an exostosis, diagnosis of ringbone is not easy, for it is then a problem of detecting the presence of a ligamentous sprain, periostitis, or osteitis. The diagnostician should take note of local manifestations of hypersensitiveness, or heat if such exist, and, in addition, other conditions must be excluded before definite conclusions are possible.

In *articular* ringbone as soon as there is developed an exostosis, it occupies a position on the dorsal (anterior) part of the articulation and extends around the sides of the joint.

Periarticular ringbone is characterized by exostoses which are situated on the sides of the phalanges and not extending around to the anterior part of the joint. This type of ringbone as well as the articular may occur "high" or "low."

Fig. 19—Phalangeal exostoses in chronic ringbone. Museum specimen of the Kansas City Veterinary College.

With the *traumatic* form of ringbone, all consequences, as to the size and form the exostosis is to assume, depend upon the nature and extent of the injury.

Rachitic ringbone is frequently observed in some sections of the country and does not ordinarily cause much if any lameness. It is a

172

disease of colts and may affect one or all of the phalanges at the same time. As the subject advances in age there is more or less diminution in the size of the enlargements.

Treatment.—Rest is essential in the treatment of ringbone. If diagnosed during its incipiency, remedial measures such as are usually employed to treat sprains, are indicated and later the parts should be blistered. When an exostosis has developed puncture firing is the remedy *par excellence*. Not that this method of treatment is infallible, for to any thinking one who takes into consideration the pathological anatomy of this condition, it is evident that no manner of treatment is beneficial in some cases. If the exostosis is so situated that it does not mechanically interfere with function, and is not so large that it may inhibit flexion and extension, and where the articular portions of the joint are not eroded, good results attend the use of the actual cautery.

In firing, after having anesthetized the extremity, and prepared the surgical area, the cautery is deeply inserted in numerous places, taking care, however, not to open the joint. The parts are immediately covered with aseptic absorbent cotton and this dressing is left in position for forty-eight hours and if perchance there is evidence of synovial discharge, the parts are again aseptically dressed in order to prevent infection of the articulation. If, as is the case usually, no perforation of the joint capsule exists, the openings made by the cautery have been closed by the coagulation of serum and there is then little chance of infection causing trouble, even though the member is left unbandaged.

In several instances, the author has treated ringbone by this method where the periarticular type existed and lameness was marked, and in three weeks the subjects were in service and not lame—this, in one instance in a valuable polo pony where the subject continued in service for more than a year without any evidence of recurrence of the lameness. The production of a deep-seated and acute inflammation with the actual cautery is preferable to any sort of counter-irritation which may be produced by vesicants.

There is no occasion for any difference in the treatment of either of the first three classes of ringbone, but in the rachitic type where treatment is given, the application of a vesicant is all that is required. In most instances treatment is not necessary.

The affected animals require a month to three months' time for recovery to take place in the average favorable cases of ringbone.

Median neurectomy is of service in many instances where lameness is not completely relieved by the use of the actual cautery and no bad results attend the performance of this operation even though no benefit is derived thereby. Plantar neurectomy is contraindicated in all cases where there exists much lameness. If lameness is due to acute inflammation bad results such as sloughing and loss of the hoof may follow; and if large exostoses mechanically interfere with function of the joint, or where articular erosions exist, no possible good can come from neurectomy. Careful discrimination should be employed in selecting cases for neurectomy for this operation; otherwise, it is very likely to prove disappointing.

Open Sheath of the Flexors of the Phalanges.

This condition does not differ from a like affection involving other tendons except that the function of these tendons is such that large synovial sheaths are necessary, and when synovitis exists, the condition then becomes more serious.

Infectious synovitis involving these tendons in the fetlock region is of more frequent occurrence than a like affection of carpal or tarsal sheaths. With the exception of the extent of the involvement and distress occasioned thereby, synovitis the result of open tendon sheaths, is similar wherever it occurs.

Etiology.—The same conditions which are responsible for open fetlock joint and other wounds of the pastern region, cause open tendon sheaths of the flexor tendons.

Symptomatology.—Because of the size and extent of this sheath and the different manner in which it is opened, there is manifested dissimilar symptoms in different cases. A nail puncture which perforates the sheath in the pastern region and at the same time produces an infectious synovitis, will cause a markedly different manifestation than will a wound which freely opens the sheath above the fetlock. In the first instance, the condition is much more painful; swelling is intense in some cases; and if the subject does not possess sufficient resistance so that spontaneous resolution promptly occurs, surgical evacuation of pus is usually necessary. When these tendon sheaths are opened, there follows a reaction which is quite analogous to that which exists in arthritic synovitis, but instead of ankylosis, adhesions with thecal obliteration occur. Rarely there result cartilaginous and osseous formations.

The constitutional disturbances which characterize this condition vary with the degree of distress occasioned. As the infection is virulent and causes serious destruction of the affected parts, so does evidence of malaise and finally distress appear. Detailed discussions of symptomatology in similar conditions have heretofore been given, and further repetition is unnecessary.

Treatment.—The same general plan of treatment which is employed for handling open joint is put in practice in these cases. Following the preoperative cleansing of the external wound and adjacent surfaces, where liberal drainage exists, tincture of iodin is injected into the sheath, the parts covered with a suitable dressing powder, and the entire member is carefully dressed with cotton and bandages.

Subsequent treatment is the same as has been outlined in the discussion of open fetlock joint on page 112. The same general plan of after-care is necessary. Recovery, however, does not require so much time ordinarily, yet punctures of the sheath occasioned by nails or other small implements make for long drawn out cases of infective synovitis.

Luxation of the Fetlock Joint.

Etiology and Occurrence. — The manner of construction of the fetlock joint is such that disarticulation without irreparable injury resulting, is practically impossible. Logically, this joint in the fore legs (not so in the pelvic limbs) should disarticulate in such manner that either all of the inhibitory apparatus (flexor tendons and suspensory ligament) must rupture or a lateral luxation is necessary. Lateral disarticulation must necessarily sever the attachment of one of the common collateral ligaments. Because of the width (transverse diameter) of the articulating surfaces of this joint, lateral luxation requires a great strain; and a force that is sufficient to occasion this trauma usually causes serious additional injury. Therefore, the condition is considered one wherein prognosis is always unfavorable in so far as practical methods of treatment are concerned.

Mr. A. Barbier[22] reports a case of bilateral luxation of the fetlock joints of the hind legs in a horse. This was done in jumping, and the extensor tendon of each leg was ruptured and the anterior portion of the metatarsus was protruding through the skin. Profuse hemorrhage had taken place due to tearing of the blood vessels.

Symptomatology. — Entire luxation of this joint when present is so evident that one cannot fail to recognize the condition. Complete disarrangement of normal relation occurs and there is either a breaking down of the inhibitory apparatus, or if a lateral disarticulation exists, the normally straight line formed by the bones of the front leg, as viewed from the front or rear, is broken at the fetlock.

Often fracture of bones are concomitant and then, of course, mobility is increased and not decreased as is the case in uncomplicated luxation.

Such violence occurs at times, when this joint is disarticulated, that the joint capsule is also completely ruptured and the articular portion of the bones is exposed to view.

Treatment. — The condition being practically a hopeless one, destruction of the subject is the thing which should be promptly done. In valuable breeding animals, owners may prefer that treatment be attempted when a lateral luxation and detachment of but one common ligament have permitted luxation without complete disarticulation and rupture of the joint capsule. In such cases, by immobilizing the affected parts as in fracture, and confining the subject in a sling for about sixty days, partial recovery may occur in some instances.

Experience has shown that where luxation with detachment of a collateral ligament occurs, recovery is slow and incomplete—there always results considerable exostosis at the site of injury.

Sesamoiditis.

Etiology and Occurrence.—Inflammation of the proximal sesamoid bones is caused by any kind of irritation which may involve this part of the inhibitory apparatus. Positioned as they are, between the bifurcations of the suspensory ligament and the pastern joint, they serve as fulcra and effectively assist in minimizing concussion which is received by the suspensory ligament. The flexor tendons also, in contracting, exert strain upon the inter-sesamoidean ligament, which has a similar effect upon the sesamoid bones as that which is produced by the suspensory ligament.

The condition occurs quite frequently, and because of the important function performed by these bones, active inflammation of the sesamoids constitutes a serious affection. Because of the fact that these bones have proportionately large articular surfaces, when they are inflamed to the extent that degenerative changes affect the articular cartilage, complete recovery seldom results.

The same pathological changes occur here that are to be seen in any case of arthritis. No special pathological condition characterizes sesamoiditis but this condition causes incurable lameness when the sesamoid bones are much inflamed.

Symptomatology.—In acute inflammation, there exist all the symptoms which portray any arthritic inflammation of like character. The parts are readily palpable and are found to be hot, supersensitive, and more or less infiltration of the tissues contiguous to the joint causes swelling. There is volar flexion of the phalanges when the subject is at rest. Lameness is intense; in some acute inflammatory disturbances the subject is unable to bear weight on the affected member.

In chronic sesamoiditis, constant lameness is the one salient feature which marks the condition. While it is possible for one sesamoid bone to become involved without its fellow being affected, this is not usual. Considerable organization of tissue surrounding the joint is present and no particular evidence of supersensitiveness exists. However, supporting weight brings sufficient pressure to bear upon the inflamed and more or less eroded bones so that pain is occasioned and lameness results.

Treatment.—During acute inflammation, absolute quiet is, of course, of first consideration. Cold packs are to be kept in contact with the parts until acute inflammatory symptoms subside. The fetlock region is then enveloped with a poultice or an iodin and glycerin combination (iodin one part to seven parts of glycerin) is applied and a dressing of cotton is kept in contact with the inflamed region. Following this, a vesicant is employed and the subject is allowed a month's rest.

In sub-acute cases, the entire region surrounding the pastern is blistered or the actual cautery is used. Line-firing is preferable. The subject is given a month or six weeks rest and one may be guided by the presence or absence of lameness as to whether improvement or recovery is taking place.

Old chronic cases, and particularly those where there are considerable induration and fibrous organization of tissue surrounding the joint, are not to be benefited by treatment.

The chief consideration in handling sesamoiditis is checking inflammation as early as possible and preventing, if this can be done, the erosion of articular surfaces. If destruction of any part of the articular surfaces can be prevented and the patient allowed ample time for complete resolution of the affected parts to occur, permanent relief is possible.

Fracture of the Proximal Sesamoids.

Etiology and Occurrence.—Fracture of the proximal sesamoid bones is caused by violent strain when there exists *fragilitas osseum*, or by contusions. The author treated a case where fracture of one sesamoid was occasioned by a horse receiving a puncture wound wherein the sharp end of a steel bar was protruding from the ground where it was firmly embedded. The subject in this case was injured while being driven along a country road. Frost[23] reports simultaneous fracture of all of the proximal sesamoids occurring in a sixteen-year-old pony. The condition is of rather common occurrence in some countries because of the fragile condition of horses' bones.

Symptomatology.—If the parts can be examined before extravasation of blood and swelling mask the condition, crepitation may be detected. In other instances, it is possible to note a displacement of parts of the sesamoid bones—this in horizontal fracture. There occurs more or less descent of the fetlock which must not be attributed to rupture of the superficial flexor tendon (perforatus). By outlining the course of this tendon with the fingers, when it is passively tensed sufficiently to follow its course, one may exclude rupture of the superficial flexor. Finding the suspensory ligament intact from its origin to the sesamoid attachments, one may also eliminate rupture of this structure as a cause of the trouble. Needless to say, marked lameness and swelling of the fetlock soon take place. The condition is painful, and ordinarily, recovery is impossible.

Treatment.—Where treatment is attempted, immobilization as in luxation is in order. The patient's comfort is sought, and if the fractured parts can be kept in close proximity, their union may occur in time. However, chances for partial recovery (which is the best to be hoped for) are so remote that early destruction of the subject is the humane and economical thing to do.

Where treatment is instituted, it is found that there is required a long time for union of the fractured bones to occur (where union does take place) and the cost of treatment together with the uncertainty of even partial recovery, makes for an unfavorable outcome. When the best possible results succeed treatment, a large callosity is formed and movement of the pastern joint is restricted. Lameness, though not intense, in the case referred to, where one bone was broken, was permanent and the subject was out of service for nearly a year.

Inflammation of the Posterior Ligaments of the Pastern (Proximal Interphalangeal) Joint.

Anatomy. — The ligaments here involved are the four volar ligaments described by Sisson[24] as follows: "The *volar ligaments* (Ligg Volaria) consist of a central pair and a lateral and medial bands which are attached below to the posterior margin of the proximal end of the second phalanx and its complementary fibro-cartilage. The lateral and medial ligaments are attached above to the middle of the borders of the first phalanx, the central pair lower down and on the margin of the triangular rough area."

This portion of the inhibitory apparatus is described by Strangeways' Anatomy as two posterior ligaments which run each from three points on the sides of the os suffraginis to a piece of fibro cartilage, described as the glenoid cartilage, and attached to the postero-superior edge of the os coronae; between them is the insertion of the inferior sesamoidean ligament.

Etiology and Occurrence. — Everything tending to increase strain upon these ligaments is contributory to possible fibrillary fracture of these structures. Excessive leverage as furnished by long toes, long toe-calks and low heels increases the normal tension on the posterior ligaments of the pastern joint. Faulty conformation, which throws an abnormal strain on these ligaments, is a predisposing cause of inflammation of these structures. Hard pulling upon slippery and rough or frozen roads is a common exciting cause of this injury. The condition is of comparatively frequent occurrence and is seen affecting draft horses frequently, in the hind legs.

Symptomatology. — Lameness is the first manifestation of this affection and weight bearing is painful in direct proportion to the extent of injury present. Volar flexion of the phalanges relieves tension on the parts; therefore, this position is assumed while the subject is at rest. When considerable tissue has been ruptured, and the condition is very painful, the foot is held off the ground as in all painful affections of the extremity.

By palpation evidence of pain is discernible, though very little swelling occurs. Pain is increased by manual tension of the parts which is done by grasping the toe of the foot and exerting traction on the flexor apparatus. Care must be taken in executing such manipulations, and it is only by comparison of the affected member with the sound one and noting the difference in the manifestations of discomfort that we may arrive at the proper conclusion.

Some hyperthermia is to be recognized in acute inflammation, by comparing the extremities. In the fore legs, navicular disease is differentiated by noting absence of contraction at the heel. By use of the hoof testers one may recognize evidence of inflammation of the navicular apparatus. In inflammation of the posterior ligaments of the pastern joint, there is also absence of the characteristic stumbling which is seen in navicular disease.

Treatment.—Rest is the first requisite, and in addition every mechanical means possible to change the center of gravity in the phalangeal region, is to be employed. This is best accomplished by shortening the toe and paring the sole at the toe as much as conditions will permit. The heel is raised by means of a shoe with moderately high heel calks.

The iodin-glycerin combination heretofore mentioned may be applied and the parts covered with cotton and bandage. Subjects require from three weeks to several months' rest and must be returned to work carefully, lest the incompletely regenerated tissues suffer injury.

Regeneration of tissue in such cases, as has been pointed out, is slow and sufficient time for complete recovery must be allowed or relapses will occur.

Fracture of the First and Second Phalanges.

Etiology and Occurrence.—Fractures of the first phalanx (suffraginis) occur with respect to frequency, second to pelvic fractures. Often, almost insignificant injuries cause phalangeal fractures. On city streets, horses shod with shoes having long calks get caught in frogs of street railways or by slipping on rails, and phalangeal bones are often broken. The author observed a case of comminuted fracture of both the first and second phalanges (suffraginis and corona) in a polo pony caused by making a sudden turn while in action in a contest on the turf.

Symptomatology.—Fracture of the phalanges is nearly always signalized by lameness, and this is marked during the period of weight bearing. Lameness is usually intense and where the pathognomonic symptom (crepitation) is not recognized, the intensity of the claudication, when other causes are absent, is indicative of fracture. The subject does not bear weight upon the affected member and where pain is intense, the foot is held in an elevated position and swung back and forth. In hind legs the member is often flexed in abduction and held in this position for several minutes, being rested on the ground only during short intervals. When compelled to walk, if pain is excruciating, the animal hops with the sound leg, no weight being supported by the fractured member.

When an examination of the subject is possible before the extremity is swollen, crepitation is usually found without great difficulty, except in a subperiosteal break or in some cases of vertical or oblique fracture. Great care is necessary in handling the injured extremity in these cases, and particularly in nervous subjects or in excited animals that have been recently injured in runaways, is it necessary to be gentle in manipulating the extremity, if definite deductions are to be made. As has been mentioned in the chapter on diagnostic principles, if the condition is so painful that the subject does not relax the parts and crepitation is masked, local anesthesia is necessary. An anesthetic solution of cocain or novocain may be applied to the metacarpal or metatarsal nerves and an entirely satisfactory examination is then possible.

Passive movement of the phalanges in all directions is practised in order to produce crepitation. When rotation of the parts does not occasion crepitation, gentle flexion and extension may do so. And in many instances, considerable manipulation of the phalanges is necessary before the pathognomonic symptom is to be recognized.

In cases where crepitation is not found and lameness is pronounced, out of proportion with other possible existing causes, one may by exclusion of other causes establish a diagnosis of fracture in the course of forty-eight hours. In the meanwhile, support is given the affected member by applying an effective leather splint, so that pain may be diminished. To combat inflammation, a suitable cataplasm may be applied directly to the skin, the extremity bandaged, and the temporary immobilizing appliance may be secured over all. In this manner one may make repeated examinations of the subject, and if slings are used and every other necessary precaution taken to promote comfort for the subject, no harm will result in delaying for several days the application of permanent immobilization—bandages and splints or casts. In fact, where much swelling exists at the time one is called to treat such cases, it is advisable to delay the application of a permanent dressing or cast until inflammation has somewhat subsided.

Course and Prognosis.—Where conditions are favorable, the nature of the fracture one that will yield to treatment, the subject not aged, and facilities for giving good attention to the affected animal are ample, fractures of the first and second phalanges recover completely in from six weeks to four months. Only simple fractures are considered curable from a practical and economical point of view, excepting in foals, where compound, and even comminuted, fractures may be so handled that animals may eventually become serviceable though blemished.

Age retards the process of osseous regeneration, but in one instance at the Kansas City Veterinary College, a very aged mare suffering from a multiple fracture of the first phalanx was treated and at the end of sixty days was able to walk into an ambulance. Large exostoses had developed and the subject remained lame, but union of the broken bone took place in a surprisingly prompt and effective manner, when age of the subject and nature of the fracture are considered.

As a rule, one is loath to recommend treatment, even in a simple transverse fracture of the first phalanx, in animals ten years of age or older. The conditions which exist in any given locality that regulate the expense of caring for an animal during the period of treatment, especially influence the course to be pursued in treating fractures.

Treatment.—For permanent immobilization of the phalanges in fracture, materials which might adapt themselves to the irregular contour of the member and at the same time contribute sufficient rigidity to the parts without doing injury to the soft structures, would constitute ideal means of treatment; but no such materials have yet been devised, and opinions are various as to the most efficient and practical method to employ.

After the fetlock has been shorn of hair and the ergot trimmed, the skin is thoroughly cleansed and allowed to dry. Several thin layers of long fiber cotton are then wrapped around the extremity—enough to pad well the member—and this is retained in position with a wide bandage. Gauze bandages are preferable to heavier bandages of cotton fabric because they are somewhat more elastic and yield to the irregular contour of the parts to a better advantage. Layers of three inch gauze bandages, which are soaked with a cold starch paste are wound about the extremity. Strips of leather that are flexible and not more than an inch in width are placed in a vertical position around the leg and these are also covered with the starch and securely held in position with the bandages. In this way, one is able to provide a sufficient degree of rigidity and at the same time, where the cast is carefully applied, little if any injury is done the skin. Such a cast is not difficult to remove and is so inexpensive that it may be removed and reapplied at any time it should be thought preferable to do so. Of course, this does not constitute an effective means of support if the parts are to be frequently and thoroughly soaked with water, but animals undergoing this sort of treatment are usually kept sheltered.

The same after-care is necessary in such cases as is given in fractures of other bones. Two months after the injury has been done, the application of a blistering ointment to the entire region is of benefit.

Results.—Much depends on the nature of fractures as to the success one may attain in approximating the parts of a broken bone, and in some cases of oblique fracture for instance, complete recovery is impossible, despite the most skillful and painstaking attention given. On the other hand, cases of simple transverse fractures make perfect recoveries in some instances. All fractures are serious, and in every instance the practitioner would best be careful to impress his client with the many difficulties which usually attend the treatment of fracture in horses.

Tendinitis.

Inflammation of the Flexor Tendons.

One of the most common causes of lameness in light harness and saddle horses is tendinitis, and because of the character of the structure of tendons and because of their function, an active inflammation of these parts is always serious.

Being almost inelastic and not well supplied with blood, tendinous tissue is slowly regenerated, and so much time is required for complete recovery to take place in tendinitis, that affected animals seldom fully recover before they are in service or vigorously exercising at will. As a result, complete recovery is delayed or prevented.

The extensor tendons, because of the nature of their function, are very seldom strained; they are often bruised and occasionally divided, but unlike this condition in the flexors, tendinitis of the extensors is of rare occurrence.

For a concise discussion of this subject the most practical classification is one made on a chronological basis and we may then consider tendinitis as *acute* and *chronic*.

ACUTE TENDINITIS.

Etiology and Occurrence.—Causes of tendinitis, as in almost all diseases, may be considered under the heads of predisposing and exciting. Among the predisposing causes of tendinitis may be mentioned, faulty conformation. Everything which has to do with increasing the strain upon tendons adds to the probability of their being over-taxed. Long, sloping, pastern bones; disproportionate development of parts, such as a heavy body and small, weak tendons and long hoofs, are the principal factors which usually predispose to tendinous sprains. Degenerative changes which take place in tendons following constitutional diseases such as influenza may also be classed as a predisposing cause.

Excessive strain when put upon tendons in any possible manner, such as is occasioned in running and jumping; making missteps and catching up the weight of the body with one foot, when the force thus thrown upon the supporting structure is great because of momentum gained at a rapid pace, are exciting causes of tendinitis.

Symptomatology.—In all cases of acute tendinitis there is presented a characteristic attitude by the subject. Volar flexion in a sufficient degree to relax the inflamed structures is always evident. The foot may be rested on the toe or placed slightly in advance of the one supporting weight, but the fetlock is always thrown forward. More or less swelling of the inflamed tendons is present. Where the deep flexor (perforans) is involved swelling is marked and with swelling there is present the other symptoms of inflammation—heat and supersensitiveness.

In manipulating tendons for the purpose of detecting supersensitiveness, care must be taken so that no false conclusion be drawn, because of the aversion many horses have to submitting to palpation of the tendons even when they are in a normal condition.

Supporting-leg-lameness is present and varies in degree with the intensity of the pain caused by weight bearing. In many instances, as soon as the subject has traveled a considerable distance, lameness diminishes or discontinues. As soon as the affected animal is permitted to stand long enough to "cool out" there is a return of the lameness, which is then marked.

No difficulty is encountered in making a practical diagnosis in tendinitis; that is, one may fail to readily recognize the extent of the involvement as it affects the superficial flexor tendon, for instance, but this has no practical bearing on the prognosis and treatment, when existing inflammation of the deep flexor is recognized.

The course of each tendon is readily outlined by palpation; all parts are easily manipulated; and with experience one may readily recognize the extent and degree of the inflammation.

Treatment.—In some cases of acute tendinitis, pain is intense and the application of cold packs during this stage is very beneficial in that pain is controlled and inflammation subsides. The extremity may be bandaged with a liberal quantity of absorbent cotton or with woolen material. Ice water is then poured around the bandaged member every fifteen minutes and this should be continued for about forty-eight hours. In some cases this treatment is not necessary for more than twelve hours; at the end of this length of time, pain has subsided and the acute stage of inflammation has passed or its intensity has been diminished.

Following the application of cold packs, the use of a poultice such as some of the sterile, medicated muds, is of marked benefit. The author has made use of tincture of iodin and glycerin in the proportion of one part of iodin to seven parts glycerin, with very satisfactory results. This combination is hygroscopic, anodyne and antiseptic and is easily applied. A liberal quantity is directly applied all around the affected tendons and the leg covered with a heavy layer of cotton, and this is snugly held in position with bandages. The application may be used once or twice daily, or if it is thought necessary, an attendant may pour a quantity of the iodized-glycerin around the leg and under the bandage once daily without removing the cotton and bandage. Needless to say, absolute rest is imperative.

When all evidence of acute inflammation has subsided vesication is indicated. At this stage walking exercise is beneficial and the subject may be allowed the freedom of a paddock.

Some practitioners are partial to the use of the actual cautery in these cases, but it is doubtful if it is necessary to produce such a great degree of counter-irritation in cases where the subject is suffering the first attack of tendinitis.

As has been indicated, ample time should be allowed for recovery and depending upon conditions, it takes from three weeks to six months for complete recovery to become established.

Chronic Tendinitis and Contraction of the Flexor Tendons.

Etiology and Occurrence.—Acute inflammation of the flexor tendons may result in chronic tendinitis. Recurrent attacks in cases where insufficient time is allowed for complete recovery to result, is followed by chronic inflammation and hypertrophy of the tendons. Again, in subjects where conformation is faulty, no amount of care will be sufficient to prevent a recurrence of the inflammation and the condition must become chronic.

Symptomatology.—On visual examination of the subject at rest, one may note the hypertrophied condition of the affected tendons. Their transverse diameter is usually perceptibly increased and in many cases, there is an increase in the antero-posterior diameter. The latter condition causes a bulging of the tendon that is so noticeable, because of the convexity thus formed, it is commonly known as "bowed tendon."

Fig. 20—Contraction of the superficial digital flexor tendon (perforatus) of the right hind leg, due to tendinitis.

In chronic tendinitis there occurs repeated attacks of inflammation wherein lameness is pronounced and there exists in reality, at such times, acute inflammation of a hypertrophic structure, where at no time does inflammation completely subside. Therefore, in chronic tendinitis there is to be found at times the same conditions which characterize acute inflammation, except that there is usually a variance of symptoms because of the difference in the degree of inflammation and pain.

The diagnosis of contraction of tendons is an easy matter because of the fact that relations between the phalanges are constantly changed with tendinous contraction. If one bears in mind the attachments and function of the digital flexors, no difficulty is encountered in recognizing contraction of either tendon.

Contraction of the superficial digital flexor (perforatus), when uncomplicated, is characterized by volar flexion of the pastern joint. The foot is flat on the ground and the heel is not raised because the superficial flexor tendon does not have its insertion to the distal phalanx (os pedis) and therefore can not affect the position of the foot.

By causing the subject to stand on the affected member, one may outline the course of the flexor tendons by palpation, and in this way recognize any lack of tenseness or contraction of tendons or of the suspensory ligament.

Fig. 21 — Contraction of the deep flexor tendon (perforans) of the
right hind leg, due to tendinitis.

Contraction of the suspensory ligament would cause the pastern
joint to assume the same position as is occasioned by contraction of
the superficial digital flexor (perforatus) tendon, but when the subject is bearing weight on the affected member, it is easy to determine that no contraction of the suspensory ligament exists, by noting an absence of abnormal tenseness of this structure. And finally,
contraction of the suspensory ligament is of rare occurrence.

Contraction of the deep flexor tendon (perforans) causes an elevation of the heel. The foot can not set flat because the insertion of the
deep flexor tendon to the solar surface of the distal phalanx (os
pedis) causes when the tendon is contracted — a rotation of the distal
phalanx on its transverse axis — hence the raised heel. No other tendon has this same effect on the distal phalanx and the condition is
correctly diagnosed without difficulty.

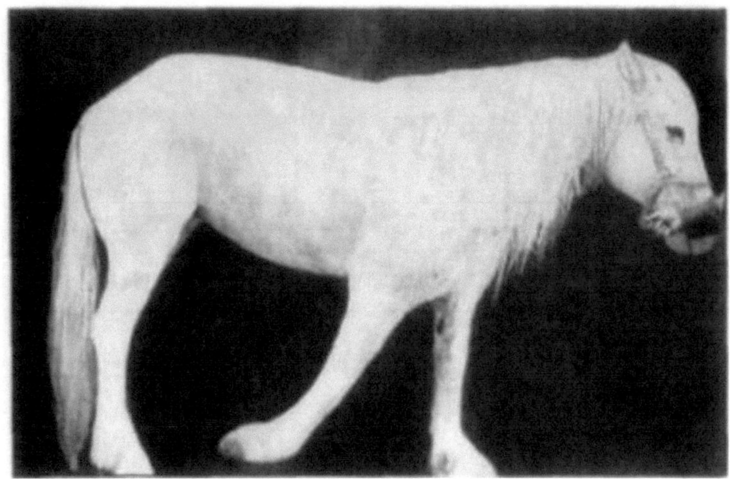

Fig. 22—A chronic case of contraction of both flexor tendons of the phalanges. In this case (presented at a clinic of the Kansas City Veterinary College) because of long continued contraction of the flexors, which prevented weight being supported with any degree of comfort, there resulted a partial paralysis of the extensors, and consequently the extremity was dragged on the ground.

Course and Complications.—This condition may exist for years without causing the subject any serious inconvenience, if the affected animal is kept at suitable work. In other instances recurrent attacks of lameness are of such frequent occurrence that the subject is not fit for service. Many affected animals that are kept in service in spite of lameness (and in some instances where no lameness is present), soon become unserviceable because of contraction of the inflamed tendon. This, in fact, is the condition which eventually becomes established in most instances.

Treatment.—Where conformation is not too faulty so that recovery may be expected, good results are obtained by line-firing the tendons and allowing the subject a few months' rest. In some cases median neurectomy is advisable. This is recommended by Breton[25] as being productive of good results even where contraction of tendons exists and tenotomy is done.

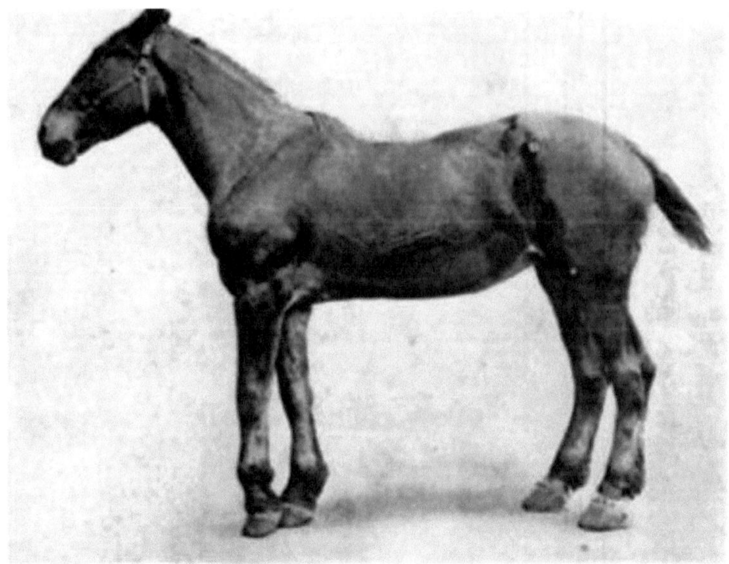

Fig. 23—Contraction of the superficial and deep flexor tendons (perforatus and perforans) of the left fore leg.

By shoeing with high heel-calks considerable strain is taken from the inflamed tendons because of the changed position of the foot which alters the distribution of weight on different parts of the leg. Rubber pads materially diminish concussion and should be made use of when the subject is returned to work, if the character of the work is such as to occasion much concussion.

It is to be remembered, however, that in sprains there occurs fibrillary fracture of soft structures and time is required for regeneration of tissue which has been injured or destroyed. Absolute rest is necessary where inflammation is acute and in sub-acute or chronic tendinitis avoidance of all work which causes irritation to the affected tendons is imperative.

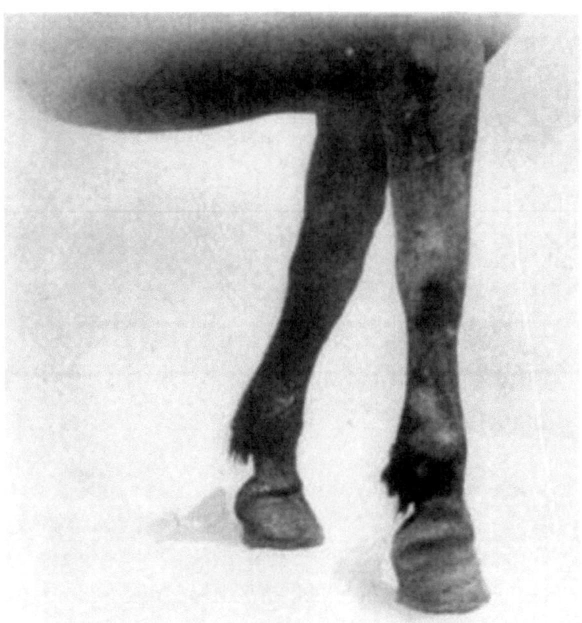

Fig. 24 — Contraction of superficial digital flexor and slight contraction of deep flexor tendon.

Where contraction of tendons exists surgical treatment is necessary. No good comes from appliances which are calculated to stretch the affected tendons; in fact, they aggravate the inflamed condition and hasten complete loss of function of the affected member. Where there exists no articular or ligamentous diseases which would defeat the purpose, tenotomy is the only remedy for contracted tendons.

Contracted Tendons of Foals.

Etiology and Occurrence.—This condition is occasionally observed and no positive explanation of the reason for its existence can be given. That mal-position *en utero* causes the metacarpal bones to develop in length so rapidly that the tendons are too short, is an explanation that is offered. Be that as it may, in breeding sections of the country the general practitioner is obliged to handle these cases and successful methods of treatment are essential even though cause is not removable.

Symptomatology.—The superficial flexor tendon (perforatus) alone, is the one usually contracted, and while both flexors are at times involved, this rarely occurs. The condition is usually bilateral.

The degree of contraction varies greatly in different cases. In some, contraction exists to such extent that it is impossible for the colt to stand, and because of continual decubitus where no relief is given, the subject is lost because of gangrenous infection occasioned by bed sores. Otherwise the same symptoms are to be observed in this condition, that exist in contraction of tendons of the mature animal.

Treatment.—Wherever contraction is not too marked and weight is borne with the affected members, and where the feet can be kept on the ground in a nearly normal position, it is possible to correct the condition without doing tenotomy. That is, in cases where the subject is simply "cock-ankled", where volar flexion of the pastern joint exists but the foot is kept flat on the ground, correction is possible without tenotomy.

In such instances the foal must be treated early—before the skin on the anterior pastern region has been badly damaged by knuckling over. It is possible in many cases to stretch the flexor tendons by grasping the colt's foot with one hand, and with the other hand one may push the pastern in the direction of dorsal flexion. This may be tried and when a reasonable amount of force is employed, no harm is done, even though no material benefit results. Some veterinarians claim good results from this treatment alone and direct their clients to repeat the stretching process several times daily.

Whether the tendons are manually stretched or not, splints should be adjusted to the affected members. The legs are padded with cotton and bandages and a suitable splint is applied on either side of the members and securely fixed in position by bandaging.

The splints are kept in position for four or five days and then removed for inspection of the affected parts. If necessary, they are reapplied and left in position for a week; however, this is unnecessary in the average case that is treated in this manner.

Where contraction exists to the extent that the subject can not stand and where no weight is borne by the feet, it is necessary to divide the affected tendons surgically. The same technic is put into practice that is employed in the mature subject but there is much greater chance for a favorable outcome in the foal. Further, if necessary, one may divide with impunity, both tendons on each leg, at the same time. In all cases this operation is done by observing strict aseptic precautions and the legs are, of course, bandaged. If both tendons are divided, splints should be employed and kept in position for ten days or two weeks. Primary union of the small surgical wound of the skin and fascia occurs in forty-eight hours.

The reader is referred to William's "Veterinary Surgical and Obstetrical Operations," for a complete description of this operation.

In veterinary literature there is occasionally described a condition which affects young foals wherein symptoms similar to those of contraction of the flexors are manifested, but upon examination it is found that rupture of the extensor of the digit (extensor pedis) exists. This affection is briefly described by Cadiot but no complete treatise on this condition has been published.

In parts of Canada foals of from one to three days of age are found affected in such manner that more or less interference with the gait is to be seen in those moderately affected. There is, in some subjects, only a slight impediment in locomotion which is occasioned by inability to properly extend the digit. In other subjects, while able to stand and walk, great difficulty is experienced because of volar flexion of the phalanges. The more seriously affected animals are unable to stand and, in most instances, perish because of the effects of prolonged decubitus.

A local enlargement occurs at the anterior carpal region and the mass is somewhat fluctuating, extravasated fluids becoming infected in many instances, and necrosis of the skin and fascia provide means for spontaneous discharge of the contents of the enlargement if it is not opened. The infection when it becomes generalized causes a fatal termination in most cases that are not treated.

Fig. 25—"Fish knees."—Photo by Thos. Millar, M.R.C.V.S.

Native stock owners of some parts of Canada know this condition as "fish knees" because of the presence of the ruptured end of the extensor tendon which is found coiled in the cavity of the enlargements caused by the ruptured tendon.

Local practitioners have treated the condition by incising the swollen mass and removing the part of tendon contained within such cavities. Treatment has not proved entirely satisfactory in the majority of instances, perhaps because of tardy interference.

In a colt's leg sent the author by Mr. Thomas Millar, M.R.C.V.S., of Asquith, Saskatchewan, a careful dissection of the carpal region revealed the fact that in this case the ruptured extensor tendon was due to injury. The colt may have been trampled upon by its dam in such manner that the tendon was divided. No noticeable evidence of injury to the skin was to be seen on its outer surface, but on the fascial side a cyanotic congested area, which was situated immediately over the site of the ruptured tendon, was very evident.

With the execution of a good surgical technic, the ruptured tendon might be sutured; the wound of the tendon sheath as well as that of the skin carefully united by means of gut sutures, the leg bandaged and immobilized with leather splints and recovery follow in a reasonable percentage of cases so treated. These cases afford an opportunity for the perfection of practical means of treatment by those who frequently meet with this affection.

Rupture of the Flexor Tendons and Suspensory Ligament.

Etiology and Occurrence. — Rupture of the flexor tendons or of the suspensory ligament is of rare occurrence. Frequently, these structures are divided as the result of wounds; but rupture, due to strain, is not frequent.

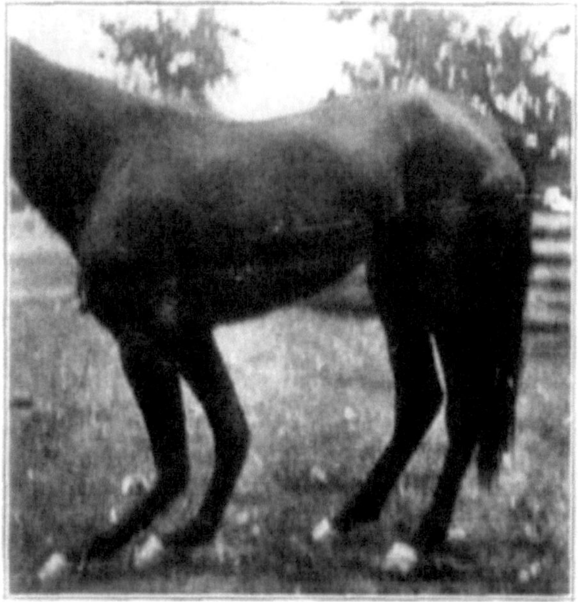

Fig. 26 — Extreme dorsal flexion said to have resulted from an attack of distemper. From Amer. J'n'l. Vet. Med., Vol. XI, No. 4.

In some cases in running horses, or in animals that are put to strenuous performances, such as are jumpers, rupture of tendons or of the suspensory ligament takes place. However, more frequently this follows certain debilitating diseases such as influenza or local infectious inflammation of the parts which results in degenerative changes and rupture follows.

The non-elastic suspensory ligament receives some heavy strains during certain attitudes which are taken by horses in hurdle jumping as is explained in detail by Montané and Bourdelle[26] under the description of this ligament. But in spite of the frequent and unusually heavy strains, which these structures receive, complete rupture is not frequently seen.

Symptomatology.—When the anatomy and function of the flexor tendons and suspensory ligament is thoroughly understood, recognition of rupture of either of these structures is easily recognized. When one considers that in rupture, a position directly opposite to that which is seen in contraction in either one of these structures, is assumed, a detailed description of each separate condition is needless repetition.

However, it is pertinent to suggest that rupture of the deep flexor tendon (perforans) allows a turning up of the toe. Whether it be torn loose from its point of attachment or ruptured at some point proximal thereto, the position is the same—heel flat on the ground, toe slightly raised and this raising of the toe varies in degree as the subject moves about.

When the superficial flexor (perforatus) is ruptured there is no change in the position of the foot but the fetlock joint is slightly lowered. The pathognomonic symptom is the lax tendon during weight bearing, which may be felt by palpation of the tendon along its course in the metacarpal region.

With complete rupture of the suspensory ligament there occurs a marked dropping of the fetlock joint and an abnormal amount of weight is then thrown upon the superficial flexor tendon (perforatus), causing it to be markedly tensed. This is readily recognized by palpation. By palpating the suspensory ligament from its proximal portion down to and beyond its bifurcation, while the affected member is supporting weight, it is possible to diagnose rupture of one of its branches.

Prognosis and Treatment.—In rupture of the superficial flexor tendon (perforatus) because of its comparatively less important function, prognosis is favorable and recovery takes place when proper treatment is put into practice.

With rupture of the deep flexor tendon (perforans), especially when it occurs at or near its point of insertion and sometimes following disease, prognosis is unfavorable.

Rupture of the suspensory ligament constitutes a condition which is, as a rule, hopeless, because of the impracticability of treating such cases.

The salient feature which characterizes any practical attempt at treatment of ruptured tendons or other portions of the inhibitory apparatus of the fetlock region, is to retain the phalanges in their normal position for a sufficient length of time that the approximated ends of ruptured tendons or ligaments may unite. The length of time required for this to occur, together with the difficulties encountered in confining the affected extremities in suitable braces or supportive appliances, precludes all possibility of this condition's being practically amenable to treatment when the deep flexor tendon (perforans) and suspensory ligament are simultaneously ruptured. It does not follow, even so, that recovery does not succeed treatment in some of these unfavorable cases.

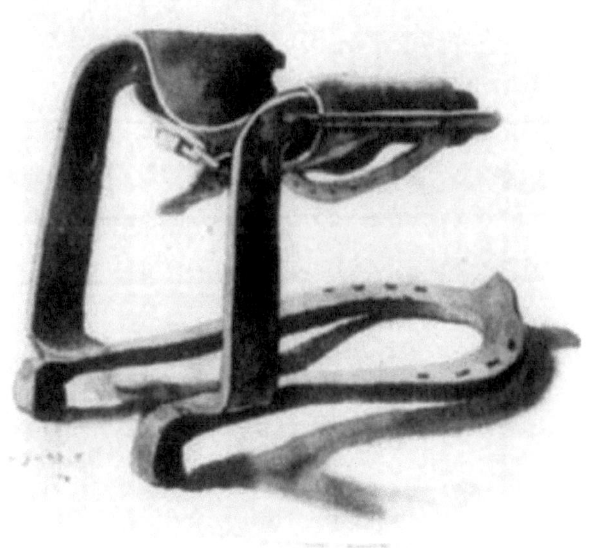

Fig. 27—A good style of shoe for bracing the fetlock where tenotomy has been performed, or in case of traumatic division of the flexor tendons. An invention of Dr. G.H. Roberts.

Affected subjects are kept in slings as long as it seems necessary—until they learn to get up without deranging the braces worn.

Several styles of braces are in use and each has its objections; nevertheless some sort of support to the affected member is necessary and steel braces which are connected with shoes are usually employed.

The principal difficulty which attends the use of braces is pressure-necrosis of the skin which is caused by the constant and firm contact of the metal support. The practitioner's ingenuity is taxed in every case to contrive practical means of padding the exposed parts in order to prevent or minimize necrosis from pressure. This is attempted—with more or less success—by frequent changing of bandages and the local application of such agents as alcohol or witch hazel. Needless to say, the skin must be kept perfectly clean and the dressings free from all irritating substances.

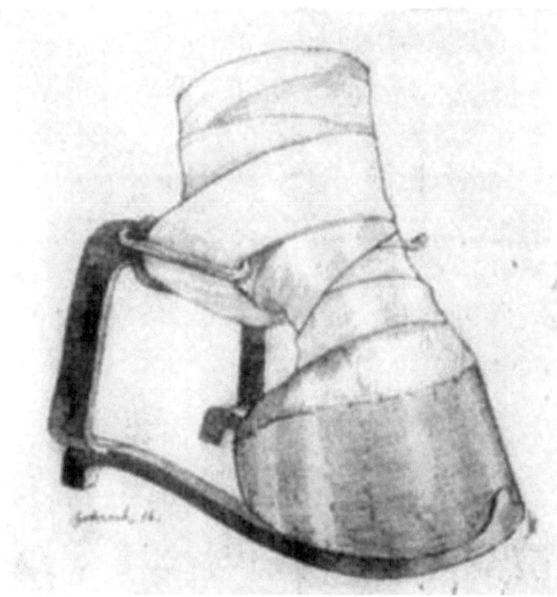

Fig. 28—Showing the Roberts brace in operation.

The fact that tendons or ligaments which are ruptured, do not re-generate as readily as in cases where traumatic or surgical division occurs, must not be lost sight of, and prognosis is given in accord-ance.

Thecitis and Bursitis in the Fetlock Region.

Etiology and Occurrence.—Synovial distension of tendon sheaths and bursae in the region of the fetlock are caused by the same active agencies which produce this condition in other parts. The fetlock region is exposed to more frequent injury than is the carpus and as a consequence is more often affected. The same proportionate amount of irritation affects this part of the leg, owing to strains, as affect the carpus from a similar cause; and synovitis from this cause, is as frequent in one case as in the other. Therefore, it is a natural sequence that the tendon sheaths of the metacarpophalangeal region are frequently distended because of chronic synovitis and thecitis. These inflammations are usually non-infective in character.

The *cul-de-sac* of the capsular ligament of the fetlock joint which extends upward between the bifurcation of the suspensory ligament is the most frequently affected structure in this region. When distended, two spheroidal masses bulge laterally and anterior to the flexor tendons in a characteristic manner. This condition is known among horsemen as "wind-gall" or "fetlock-gall."

The sheath of the flexor tendons, which begins about the middle portion of the lower third of the metacarpus, and continues downward below the pastern joint is often distended.

Excepting in cases of acute inflammation attending synovitis of these parts, no lameness marks its existence and in chronic cases of synovial distension the service of affected animals is not interfered with. These distensions constitute unsightly blemishes and they are treated chiefly for this reason.

No difficulty is encountered in recognizing these conditions even where considerable organization of tissue overlying distended thecae occurs. In such cases there may be only slight fluctuation of the enlargement, but if necessary, an aseptic exploratory puncture may be made with a suitable needle or trocar.

Treatment.—Complete rest and the local application of cold packs are in order in acute synovitis when there is distension of tendon sheaths. In the fetlock region, because of the ease with which pressure may be employed, the parts should be kept snugly wrapped with cotton, and derby bandages are used to exert the desired amount of pressure over the affected region. The pressure-bandages should be employed as soon as all acute and painful inflammation has subsided; and then they should be continued, day and night, for ten days or two weeks. The bandages should be removed morning and night. After the skin of the leg has thoroughly dried off, an infriction of alcohol or distilled extract of hamamelis is given the parts and the cotton and bandages are readjusted. A good, even and firm pressure in such cases is productive of satisfactory results.

Fig. 29—Distension of theca of the extensor of the digit (extensor pedis).

In chronic distensions of tendon sheaths synovia may be aspirated and about five cubic centimeters of equal parts of tincture of iodin and alcohol is injected into the cavity. This is not done, however, without usual aseptic precautions. If no marked swelling results within forty-eight hours the entire fetlock region is thoroughly vesicated and, as soon as the skin has recovered from the effects of the vesicant, pressure bandages may be employed. In these cases, subjects may be put into service after all swelling which the injection or the vesicant has produced has subsided. The pressure bandages are used at night or during the time that the horse is in its stall and they are not worn by the subject while at work.

Where no marked swelling occurs within ten days, as the result of the injection of iodin, the injection may be repeated and, if thought necessary, the quantity may be materially increased. If swelling does not occur it is indicative that no particular irritation has been caused.

Some swelling is desirable and much swelling sometimes results and persists for weeks. This is not in any way likely to cause permanent trouble; and if the technic of injection is skilfully executed no infection will follow.

By persistent and careful use of suitable elastic bandages, the support thus given the parts, together with the absorption of products of inflammation which constant pressure occasions, some chronic cases of synovial distension of tendon sheaths recover in two or three months and this without other treatment. Such good results are not to be expected in aged subjects, nor in horses having at the same time, chronic lymphangitis.

Where bandages of pure rubber are employed great care is necessary, if one is not experienced in their use, lest necrosis result. Where bandages are uncomfortably tight the subject will manifest discomfort, and an attendant should observe the animal at intervals for a few hours (where there may be some doubt as to the degree of pressure which is exerted by elastic bandages) and readjustment made before any harm is done.

Arthritis of the Fetlock Joint.

Anatomy. — The anatomy of the metacarpophalangeal articulation is briefly reviewed on page 58 under the heading of "Anatomo-Physiological Review of Parts of the Foreleg."

Etiology and Occurrence. — The chief causes of non-infective arthritis of the fetlock joint are irritations from concussion and contusions due to interfering. The condition occurs in young animals that are over-driven in livery service or other similar exhausting work, where they become so weary that serious injury is done these parts by striking the pasterns with the feet — interfering. In these "leg-weary" animals, that are always kept shod with fairly heavy shoes, much direct injury is done at times by concussion due to self-inflicted blows. In older animals, where there exists similar conditions, with respect to their being worn from fatigue and, in addition, periarticular inflammatory organizations, arthritis is not of uncommon occurrence.

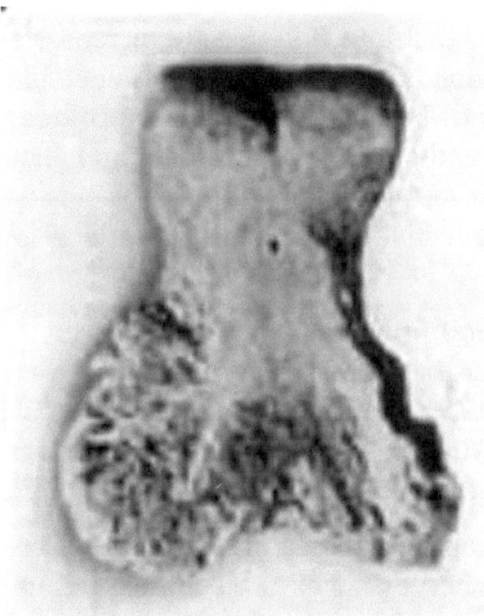

Fig. 30 — Rarefying osteitis wherein articular cartilage was destroyed in a case of arthritis of fetlock joint.

Symptomatology. — In true arthritis there exists a very painful affection which is characterized by manifestations of distress. The subject may keep the extremity moving about—where pain is great—suspended and swinging. There is swelling which is more or less hot to the touch and compression of the parts with the fingers increases pain. Lameness is always pronounced and no weight is supported with the affected member in very acute and generalized arthritic inflammations. There occurs the usual facial manifestations of pain—the tense condition of the facial muscles and the fixed eye and nostril are in evidence.

In cases where there exists a synovitis or where a very limited portion of the articulation is involved, a somewhat different clinical picture is presented. Then, the disturbance causes less distress; local swelling and evidence of supersensitiveness are not so pronounced and lameness is not intense, though weight-bearing is painful.

Prognosis.—There is a constant difference in the degree of pain manifested, as well as the other symptoms of inflammation, between true arthritis, which involves much of the joint, and synovitis; or synovitis plus a small circumscribed area of joint involvement. This difference is present in all joint affections of the extremities and, in passing, it is well to say that infection usually increases every manifestation of pain. Infection occasions more pronounced local symptoms of inflammation and, because of the rapid progress of necrotic destruction of cartilage, the course of the affection is usually rapid; ankylosis is a frequent result and loss of the subject is often inevitable. However, in non-infective arthritis of the fetlock joint, prognosis is favorable.

Treatment.—The same general principles which are employed in arthritis of other joints are used here. Rest and comfort for the patient is sought in every available manner. If the subject remains standing too long, the sling should be used and a well-bedded box-stall will contribute much to the comfort of the patient.

Pain and acute inflammation is diminished or controlled, if possible, by using ice-cold packs. In nervous, well-bred animals analgesic agents may be employed; or small doses of morphin sulphate—one to two grains—given at intervals of three hours during the first stages of the affection is very beneficial. This is especially indicated in infectious arthritis.

As inflammation subsides, hot applications are used and finally counter irritants are employed. Their selection is a matter of choice with the practitioner. The object sought is the same with every practitioner and while methods employed vary, results are not markedly different except in so far as the degree of counter irritation which is produced varies in given cases. Where a great degree of counter irritation is thought necessary, line-firing with the actual cautery is the remedy *par excellence*.

After-care. — In the course of three or four weeks subjects may be allowed the run of a paddock and, after a complete rest of six weeks at pasture, they may be returned to work with care, if the work is not of a nature to occasion concussion or other manner of irritation to the articulation.

Neurectomy is not indicated even though there is a recurrence of lameness, unless the lameness is not pronounced and inflammation is periarticular and no osseous enlargements mechanically interfere with function of the joint. There are few cases then, where neurectomy is materially helpful.

Ossification of the Cartilages of the Third Phalanx. (Ossification of the Lateral Cartilages.)

Anatomy and Function of the Cartilages.—Surmounting each wing of the distal phalanx (os pedis) is the irregularly-quadrangular cartilage. The superior border of this cartilage is thin, generally convex, and perforated for vessels to pass to the frog; the inferior border is attached to the wing of the third phalanx and posteriorly, it is reflected inward and is continuous with the inferior surface of the sensitive frog. The anterior border which is directed obliquely downward and backward becomes blended with the anterior lateral ligament of the coffin joint. The fibrous expansion of the anterior digital extensor (extensor pedis) is united to the anterior borders of the lateral cartilages.

According to Smith[27]: These structures form an elastic wall to the sensitive foot, and attachment to the vascular laminae; they also admit of increase in width occurring at the posterior part of the foot without destroying the union of the two set of leaves. Further, by their connection with the vascular system of the foot, their elastic movements materially assist the circulation. The primary use of the lateral cartilages is to render the internal foot elastic, and admit of its change in shape which occurs under the influence of the weight of the body. The alteration in the shape of the foot is brought about by pressure on the pad, which widens and in consequence presses on the bars. The pressure received by the pad is also transmitted to the plantar cushion, which likewise flattens and spreads under pressure. Both of these factors force the cartilages slightly outwards. When the posterior wall recoils the cartilages are carried back to their original position. Should the elastic cartilage under pathological conditions become converted into bone, its functions are destroyed, and lameness may occur.

Etiology and Occurrence.—The causes of ossification of these cartilages are several. No doubt there exists a predisposition to this condition for it is of such frequent occurrence in heavy draft types of horses. Concussion plays an important rôle and, according to Möller's[28] theory, which is sound, high heel calks prevent the frog from contacting the ground, and as weight is placed upon the foot "the lateral cartilages are subjected to a continuous inward and downward dragging strain."

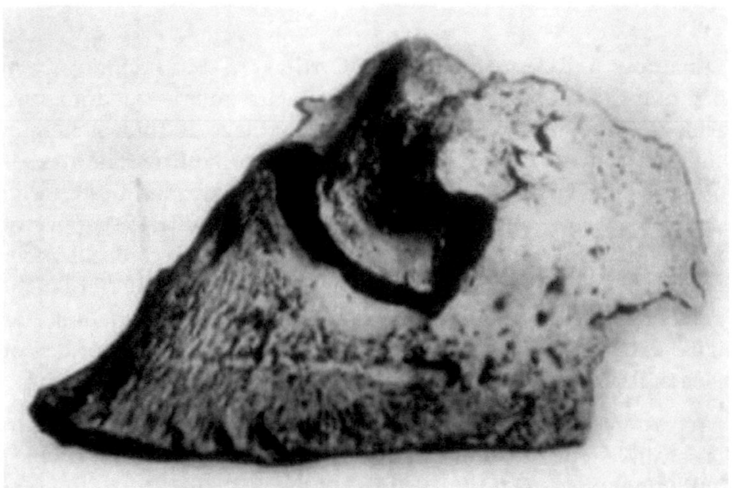

Fig. 31 — Ringbone and sidebone.

The condition affects the cartilages of the fore feet more frequently than those of the hind and the outer cartilage is more often ossified than is the inner. This fact may be accounted for by its more exposed position; it is also frequently injured by being trampled upon and otherwise contused or cut, as in lacerated wounds of the quarter.

Symptomatology.—Ossification of the cartilages is known by grasping the free borders with the fingers and attempting their flexion; the rigid inflexible ossified cartilage is thus easily recognized.

Lameness during weight-bearing occurs in the majority of cases at some time. Much depends on the conformation of the foot and whether the involvement affects one or both cartilages as to the degree and duration of lameness which marks this affection. In narrow and contracted heels it is reasonable to expect more lameness than in well formed feet. Where only one cartilage is ossified, the other being flexible, there is less inconvenience experienced by the subject during weight-bearing, because of the expansion of the heel which the one normal cartilage allows.

218

Treatment. — There is little if anything to be done in case the cartilage has become ossified except to shoe without high calks but preferably with rubber pads. The hoof should be kept moist; the wall at the quarter may be rasped thin and kept anointed. Firing is of no practical benefit in these cases, and it is doubtful if vesication is helpful excepting where only a part of the cartilage is ossified.

Subjects which continue somewhat lame, because of complete ossification of both cartilages, are best put to slow work on soft ground and not driven on pavements.

Navicular Disease.

This more or less ambiguous term has been applied to various diseases affecting the structures which make up the coffin joint. We consider this name to be applicable to inflammatory involvement of the third sesamoid (navicular bone), the deep flexor tendon (perforans) and the bursa podotrochlearis or navicular bursa.

Etiology and Occurrence.—In 1864 Thomas Greaves[29] wrote on the subject of navicular disease as follows: "The opinion I entertain upon the subject of navicular disease is, that in by far the greater majority (if not all) of these cases there exists in the animal affected a congenital tendency or predisposition, that, generally speaking, it is the high stepper, the good goer, that becomes the victim to this disease; and it is a fact well attested, that it as frequently develops itself in the feet with wide frogs, bulbous heels, shallow heels, spread flattish feet, as in the narrow upright feet.... I have known foals, born from defective parents, in which this condition was so strongly developed, that all men would at once pronounce them affected with navicular disease, and such lameness was permanent."

Often both fore feet are affected and this would point toward its being a disease wherein either conformation or congenital tendencies exists. It is rare that hind feet are involved.

There are many theories regarding the possible exciting causes of navicular disease and, when one has carefully considered the explanations as offered by Peters, Möller, Branell, Schrader and others, he may conclude that navicular disease is a non-infectuous inflammatory affection of the third sesamoid (navicular) bone, deep flexor tendon (perforans) and adjoining structures. Whether it originates in the flexor tendon or whether the bone is the original part affected, the disease is frequently met, and of all possible causes, jars and irritation incident to concussion of travel, are probably the principal causative agents.

Symptomatology.—Lameness is the primary indicator and a constant symptom which attends navicular disease wherever much structural change affects the infirm parts. As the degree of intensity or extent varies, so is there a dissimilarity in the character of the impediment. Incipient cases of bilateral involvement are more difficult to diagnose than are unilateral affections, particularly when lameness is not marked. There is manifested a supporting-leg-lameness which varies as to degree in the same subject at different times. This may be noticed during the same trip in an animal that is being driven. There is a tendency for the subject to stumble and, of course, where the affection is bilateral, there is a stilted gait owing to shortened strides.

At rest the lame animal usually points with the affected member. Because of the fact that the distance is lessened between the origin and insertion of the deep flexor tendon (perforans) by this attitude, one may readily understand the reason for the position assumed by the subject. Pressure on the navicular bone is diminished and tension on the flexor tendon is relieved by even slight volar flexion.

In acute inflammatory affections abnormal heat may be detected in the region of the heel. By exerting tension on the flexor tendon, by means of passive dorsal flexion of the member, evidence of hyperesthesia may be detected. With the hoof testers one may determine supersensitivenss in most instances. There occurs more or less contraction of the hoof in navicular disease, but this is not to be taken as a cause of the affection, but rather a sequence.

Fig. 32 — "Pointing" — the position assumed by horse having unilateral navicular disease.

In some cases of unilateral navicular disease there is a marked contrast in size between the sound and unsound foot. However, one must not be misguided in this particular, for in some pairs of sound feet there exists considerable difference in size. Finally, by a change from the normal position of the foot to one in which the heel is somewhat elevated (as may be obtained by shoeing with high heel calks), relief is evident, and in the opposite position, the condition is aggravated. This experiment may be used for diagnostic purposes.

Treatment. — When the anatomy of the diseased parts is taken into consideration, and an analysis of the lesions which occur in cases where considerable structural change is occasioned by this affection, it is obvious that recovery is impossible. Only in cases where the inflammation is promptly checked before damage has been done the navicular bone or the flexor tendon, is permanent recovery possible. The disease is not frequently treated during this stage, however, and in the majority of instances the condition becomes chronic.

As soon as a diagnosis is made the shoes must be removed, the toe shortened with the hoof pincers and rasp and the subject is put in a well bedded box-stall. If the animal is very lame and the inflammation is acute, ice-cold packs should be applied to the feet. As soon as acute inflammation has subsided the foot may be so pared that all excess of sole and frog is removed without lowering the heels, and the animal may be blistered about the coronet region. The subject may be shod later, with heel calks that raise the heel moderately and a protracted period of rest should be enforced.

In cases where no acute inflammatory condition exists, neurectomy is beneficial. One must discriminate, however, between favorable and unfavorable subjects. This is not a last resort expedient to be employed in cases where extensive lesions of the navicular structures exists. With proper shoeing, and by putting the subject at suitable work, where concussion of fast travel on hard roads is not necessary, the best results are obtainable.

Laminitis.

This disease is primarily a non-infective inflammation of the sensitive laminae which very frequently affects the front feet. Often all four feet are affected, less frequently one foot (when its fellow is unable to sustain weight) and rarely the hind feet alone.

Occurrence.—Probably a greater number of cases of laminitis occur in localities where horses that are worked on heavy transfer wagons are, when in a state of perspiration, allowed to stand exposed to sudden lowering of temperature and to stand in a cool or cold shower of rain such as occurs near the coast of the Great Lakes or the ocean in some parts of this country.

This disease occurs in connection with digestive disorders of various kinds and, because of the frequent association of the two conditions, the common term "founder" has long been employed to designate laminitis. In cases of "over-loading," particularly when a large quantity of wheat has been eaten by animals that are unaccustomed to this diet, laminitis almost constantly results.

Large draughts of cold water, when drunk by animals that are overheated is often followed by laminitis. Concussion, such as attends hard driving, especially in unshod horses or on rough and hard roads, is often succeeded by this affection. Likewise, as has been stated, injury such as is occasioned by long continued standing on the same foot is followed by laminitis. Some horses that are frequently shod, suffer from this affection a few hours after shoes have been reset. Dr. Chas. R. Treadway of Kansas City reports the rather frequent occurrence of such conditions in horses that are in the fire department service in his city.

Age in no way influences the occurrence of laminitis and the general condition of an animal with regard to its vigor or state of flesh has no apparent influence toward predisposing horses to this ailment.

Etiology and Classification.—As it is with some other diseases, one may unprofitably theorize on cause and readily enumerate many conditions which are apparently contributory toward producing the affection. Causes may well be grouped, however, and a more definite understanding of laminitis is possible as a result. Such collocation would include conditions which directly or indirectly affect the digestion, such as puerperal laminitis, drinking of large quantities of cold water and exposure to cold and rain when the body is warm. All of these various conditions might be said to affect the vaso-constrictor nerves in such manner that the natural tendency (because of the peculiar structure of the sensitive laminae and their mode of attachment to the non-sensitive wall) which solipeds have for this affection is indirectly due to this one cause—vaso-constriction. According to Dr. D.M. Campbell, the effect of toxic materials, which may be absorbed from the digestive tract or the uterus in parturient females, upon the vaso-constrictor nerves, is such that a passive congestion of the sensitive laminae occurs and laminitis is the result. He believes that even the chilling of the surface of the body when very warm, by a cold rain, constitutes a condition wherein the effect upon the vaso-constrictors is the same.

This grouping does not include the effect of direct injuries of any and all kinds to which the feet are subjected such as: Concussion in fast road work, injuries occasioned by tight or ill fitting shoes, contusions of any kind resulting in non-infectious inflammation of the sensitive laminae, as well as the causes which produce laminitis where weight is borne by one foot when its fellow is out of function.

A classification which is practical is that of *acute* and *chronic* laminitis. To the practicing veterinarian it is this manner of consideration that is essential in the handling of these cases.

Symptomatology.—In the acute attack the condition is so well described by Dr. R.C. Moore[30] that we quote him in part as follows:

The acute form is generally ushered in very suddenly. Often a horse that is perfectly free from symptoms of the disease is found a few hours later so stiff and sore that he will scarcely move. They stand like they were riveted to the ground. If forced to move the evidence of pain subsides to some extent after they have gone a short distance, to return more severe than ever after they have been allowed to stand for a short time. If the disease is confined to the two front feet, the hind feet are placed well under the center of the body to support the weight and the front ones are advanced in front of a perpendicular line so as to lessen the weight they must bear. If they are made to move, the same position of the feet is maintained. If made to turn in a small circle, they do so by using the hind feet as a pivot, bringing the front parts around by placing as little weight on them as possible.

Placing the hind feet so far under the body, arches the back and often leads to errors in diagnosis, the condition sometimes being taken for diseases of the loins or kidneys.

If all four feet are involved, the animal stands in the usual position assumed in health, but if urged to move, the least effort to do so usually brings on chronic spasms of the entire body. In very severe cases, a slight touch of the hand will develop the spasms. At times they are so severe, and have such short intermissions, that the disease has been mistaken for tetanus. However, the clonic nature of the spasm should prevent such an error. If they are lying down, it is difficult to get them to arise, and if they do so, they show marked symptoms of pain for some time after rising.

If the disease is confined to the hind feet, they are placed well forward to relieve the strain on the toe caused by the downward pull of the perforans (deep flexor) tendon, but in place of the front feet being kept in front of a perpendicular line, as they are when the disease is confined to the front ones, they are placed far back under the body, so they will carry the maximum share of the body weight of which they are capable. The position of the feet is of great importance and offers symptoms that should not be overlooked.

When the subject is caused to walk, symptoms of excruciating pain are manifested in all acute cases of laminitis. In some cases where all four feet are affected, no reasonable amount of persuasion will cause the suffering animal to move from its tracks.

There is acceleration of the rate of heart action; the pulse is full and in some cases, bounding. As the affection progresses the pulse becomes rather weak and irregular. The character of the pulse in the region of the extremity is a reliable indicator; but one has to learn to make necessary discrimination because of the condition of the parts, as in some cases of lymphangitis or where the skin is abnormally thick. The characteristic throbbing pulse is, however, easily recognized in most cases. Temperature is variable, though usually elevated from one to four degrees above normal. This symptom varies with the type and stage of the affection. In a subject that has been down, unable to rise for several days, where there is a suppurative and sloughing condition of the laminae, the temperature is high. Whereas, in some other and less destructive cases there may be little thermic disturbance after the first few hours have lapsed.

A constant symptom in bilateral affections of acute laminitis is the difficulty with which the subject supports weight with one foot. It is this which causes the victim to stand as if "rooted to the ground" when all four feet are involved. If one attempts to take up one foot, thus causing the subject to stand on the other, there is much resistance and in many cases the animal refuses to give the foot.

When we consider that the sensitive parts of the foot are encased by a horny, unyielding box and that, when the laminae are congested, a great pressure is brought to bear upon the sensitive structures, it is easy to understand why the condition is so painful.

Chronic laminitis is a sequel of acute inflammation of the sensitive laminae. It varies as to intensity and the exact manner of its manifestation depends upon preëxisting disturbances.

In some mild cases of laminitis there are recurrent attacks wherein no particular structural change exists, and diagnosis is established chiefly by noting the character of the pulse at the bifurcation of the large metacarpal (or metatarsal) artery just above the fetlock. The same manifestation of pain is present when weight is supported by one foot, though in a lesser degree. There is less local heat to be detected by palpation than in the acute cases.

Chronic laminitis as it occurs following acute attacks which have resulted in structural changes of the foot, present the same symptoms just described and, in addition, the peculiar alterations in structure exist. When, owing to acute inflammation of the sensitive laminae, there has resulted necrosis of this sensitive tissue together with infiltration between the anterior surface of the distal phalanx (os pedis) and the contacting hoof, the lower portion of the distal phalanx is turned downward and backward (rotated upon its transverse axis). Because of the traction which is exerted by the deep flexor tendon (perforans), as it attaches to the solar surface of the distal phalanx, this rotation is facilitated. With hyperplasia of lamina, at the anterior portion of the distal phalanx, there results a thick "white line." Rotation of the distal phalanx necessitates a descent of its apical portion and there occurs a "dropped sole."

In time, partly because of excessive wear of hoof at the heel, owing to an altered condition in the normal antagonistic relation between the flexor and extensor tendons, the toe makes an excessive growth, and the concavity of the anterior line is accentuated owing to this abnormal length of hoof. The hoof, because of recurrent inflammatory attacks, is corrugated—elevations of horn in parallel rings are usually present.

Fig. 33—The hoof in chronic laminitis. Note the concavity. This animal was serviceable for any work that could be performed at a walk.

Animals that are so affected in traveling strike the heel first and the toe is later contacted with the ground surface. Rotation of the distal phalanx upon its transverse axis produces a condition, with respect to this peculiar impediment, that is equivalent to added and excessive length of the deep flexor tendon.

Where there occurs suppuration, by careful inspection of the coronary region, one may early recognize detachment of hoof. In such cases animals remain recumbent and, while the condition is not so painful at this stage, the practitioner must not overlook the real state of affairs. History, if obtainable, will be a helpful guide in such cases. Separation of hoof occurs as a rule in from four to ten days after the initial attack of acute laminitis. Needless to say these cases are hopeless, when the economic phase of handling subjects is considered.

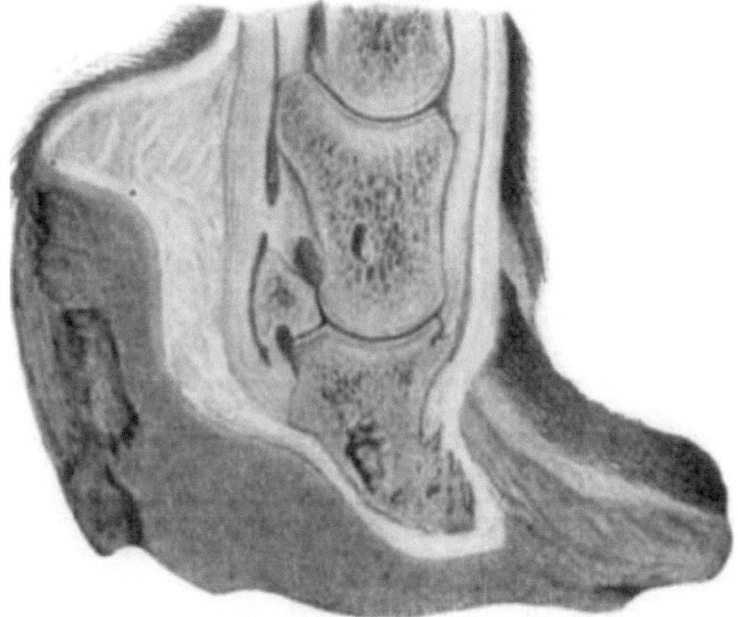

Fig. 34—Showing the effects of laminitis. By permission, from Merillat's "Veterinary Surgical Operations."

Treatment.—Much depends upon the concomitant disturbances (or causes if one is justified in referring to them as such) as to the manner in which laminitis is to be treated. In all cases where digestive disturbances exist, the prompt unloading of the contents of the alimentary canal is certainly indicated. D.M. Campbell[31] in a discussion of laminitis has the following to say regarding the treatment of such cases:

Because superpurgation may be followed by laminitis, the advisability of using the active hypodermic cathartics is questioned. Neither arecolin nor eserin can cause superpurgation. The action of the former does not continue longer than an hour after administration and of the latter not more than eight hours. The action of either is mild after the first few minutes.

I do not think that anyone has recommended either arecolin or eserin where there is severe purgation. Where the intestinal canal is fairly well emptied and its contents fluid, I should be inclined to rely upon intestinal antiseptics to hold in check harmful bacterial growth.

The use of alum in the treatment of laminitis is held to be without reason other than the empirical one that it is beneficial. If laminitis is due chiefly to an autointoxication, good and sufficient reason for the administration of alum can be shown based upon its known physiological action. It is the most powerful intestinal astringent that I know of and has the fewest disadvantages. I have not noted constipation following its use nor diarrhea, nor a stopping of peristalsis, nor indigestion, and in any case its action lasts at most only a few hours, and if it did all these, it could not much matter. Quitman says, that it constricts the capillaries. If this is true, a thing of which I am not certain, is it not reasonable to suppose that as with other vaso-constrictors, e.g., digitalis, there is a selective action on the part of the capillaries (not of the drug) and those that need it most, i.e., those of the affected feet in laminitis, are constricted most? All body cells exert this selective action in the assimilation of food, the tissue needing most any particular kind of food circulating in the blood, gets it.

Our first consideration in laminitis should be to remove the cause — to stop the absorption of the toxin in the intestinal tract that is producing the condition. This we accomplish by partially unloading it by the use of the active hypodermic cathartics and stopping absorption by the surest and most harmless of intestinal astringents. Whether the astonishingly prompt and certain action of alum in this case is due wholly to its astringent action or whether alum combines with the harmful bacterial products chemically and forms an innocuous combination, I can only surmise, and it is unimportant. At any rate, when alum is administered, the onslaught of the disease is promptly stopped. Irreparable damage may already have been done if the case is a neglected one, but whether administered early or late in acute attacks, the progress of the disease is stopped immediately.

The same authority may be profitably quoted in the matter of handling all cases wherein the revulsive effect of agents which diminish vascular tension are chiefly indicated or necessary as adjuvants. In this connection, Campbell says:

The early and vigorous administration of aconitin in laminitis to its full physiological effect, is more logical. Assuming that laminitis is due to absorption of harmful products from the intestinal tract permitted through the deranged functioning of the organs of digestion, or assuming that it is due to an extension of the inflammation from the mucosa to the sensitive lamina, or that it is a reflex from a sudden chilling of the skin, we have in any of these conditions a disturbed circulation, and aconitin is the first and foremost of circulation "equalizers." Furthermore, in laminitis there is an elevation of the temperature, an almost invariable indication for aconitin. A speedy return of the temperature to normal, a very marked diminution of the pain and improved conditions generally, appear coincident with the symptoms of full physiological effect of aconitin when given in cases of laminitis, which constitutes assuredly an important part of its treatment.

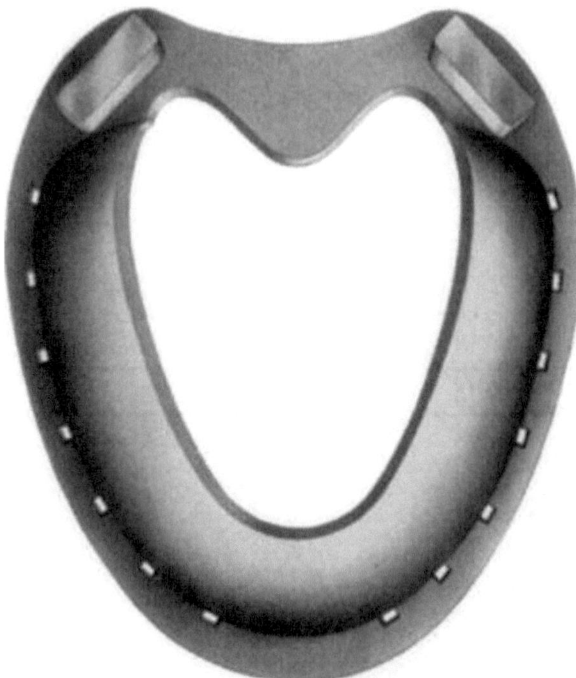

Fig. 35 — Inferior (convex) surface of Cochran shoe.

Where lameness is not great as in cases wherein no marked structural change of the foot has occurred, proper shoeing is very beneficial. By keeping the heels as low as possible and shoeing without heel calks a more comfortable position is made possible. Thin rubber pads which do not elevate the heel are of service in diminishing concussion.

Dr. David W. Cochran of New York City has attained unusual success in cases of chronic laminitis with dropped sole by the use of a specially designed shoe.

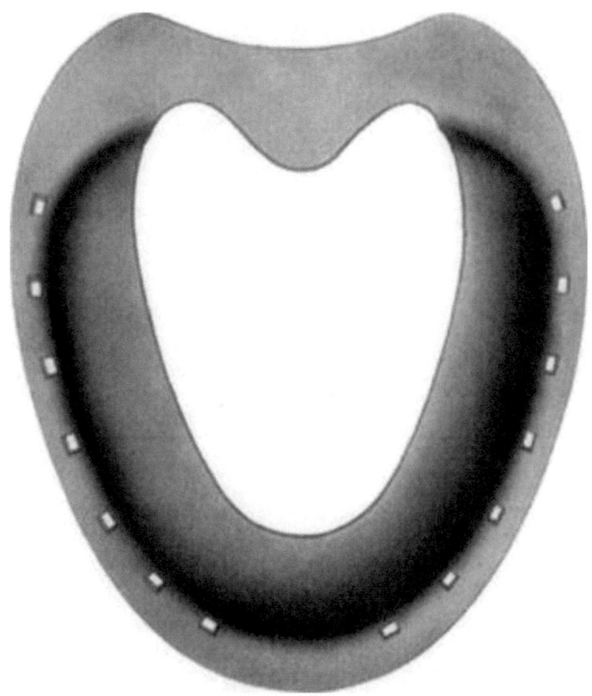

Fig. 36—Superior surface, showing concavity or bowl, as formed by the toe and branches of the shoe, as designed by Dr. David W. Cochran.

Cochran claims that, not only are horses with dropped soles that would otherwise have to be put off the streets enabled to do a fair amount of work by means of this shoe, but that continually wearing it, meanwhile keeping the convexity of the front of the hoof rasped thin, in time brings about a marked improvement, and that after some months or years of use the animals are able to work with ordinary rubber-pad shoes, provided they are arranged to facilitate breaking over.

From having been successfully used on some race horses of high value, the Cochran shoe has attained considerable notoriety and is being used by a number of practitioners. A disadvantage, however, arises from the fact that few horseshoers other than Doctor Cochran seem able to make the shoe, the peculiar shape of which offers considerable difficulty in forging. Concerning the application of the shoe Cochran[32] says:

"The most important primary procedure is the preparation of the foot to receive the shoe. All excess of growth must be removed from the anterior face of the hoof. The outer face must be reduced at the toe (not shortened), but rasped down thin for the lighter the top of the foot is, the more chance the sole and coffin bone will have of resuming their former normal position. The pressure of the wall at the toe upon the exudate between wall and coffin bone, tends to force the coffin bone and sole out of their normal position. Leave the sole alone. You can lower the excess of growth at the heels.

"There are many designs of shoes to relieve this condition. A great deal depends on the judgment of the shoer to meet the conditions presented, depending on the degree of the convexity and strength of the sole. In some cases we use a shoe that admits of a large amount of sole room. Again, we shoe with a shoe of wide cover. In other cases a shoe with even pressure over the whole sole. In some cases a high, narrow shoe, resting only on the wall, or the ordinary plain shoe with side calks welded close to the outside edge and the shoe dished well from these as a foundation. Then we have the air cushion pad designed after the model of the bowl shoe."

In cases when slight and persistent lameness interferes sufficiently to prevent using an animal at any sort of work on hard roads, median neurectomy will relieve all lameness in most instances. This is a safe operation, moreover, in that no bad after effects are to be feared, even though lameness were to continue.

Calk Wounds. (Paronychia.)

Etiology and Occurrence.—Injuries of various kinds are inflicted upon the coronary region but usually they are due to the foot being trampled upon. When the foot that inflicts the injury happens to be unshod, a contusion of the injured member is occasioned, but in the majority of instances, wounds that demand attention are the result of shoe calks which have penetrated the tissues in the region of the coronary band. Often calk wounds are self-inflicted. When animals are excited and in turning crowd one another, they often perform dancing movements which frequently result in deep calk wounds of the coronet. Some horses have a habit of resting the heel of one hind foot upon the anterior coronary region of the other. While sleeping in this position, if they are suddenly awakened, the weight is abruptly shifted to the uppermost foot and the one underneath is (because of the pain attending its being wounded) quickly drawn out from under its fellow. In this way deep cuts may divide the coronary band and inflict extensive injury to the sensitive lamina as well.

An infectious type of coronary inflammation occurs in some localities during the winter months, wherein the condition is enzootic.

Symptomatology.—Depending upon the manner in which the injury has been produced, the appearance of the wound varies and likewise lameness is more or less pronounced. If the tissues are not divided and the wound is chiefly of the subsurface structures, there will not immediately occur pronounced local evidence of the existence of injury; but as soon as the lame animal is made to move, the peculiar character of the impediment (supporting-leg lameness with the affected foot kept well in advance of its normal position) directs attention to the extremity and all of the symptoms of acute inflammation are discovered.

Where a wound is inflicted which divides, in some manner, the surface structures (skin, coronary band, or the hoof wall) one's attention is at once called to the existence of the wound.

Because of the fact that there is every facility for the production of a sub-coronary and podophylous infection, these wounds should receive prompt attention. In some instances, the pastern joint is opened by calk wounds and then, of course, an infectious arthritis succeeds the injury.

Treatment. — In all contused wounds of the coronary region the parts need thorough cleansing; the hair, if long is clipped and a cataplasm is applied. Or preferably, an iodin-glycerin combination of one part of iodin to four parts of glycerin is poured on a layer of cotton, and this is confined in contact with the inflamed parts by means of a bandage.

Where normal resistance to infection obtains, the subject usually suffers no suppurative disturbance when the surface structures are not broken; and daily applications of the antiseptic lotion above referred to stimulates complete resolution. This may be expected in from four to ten days depending upon the extent of the injury.

If a calk wound has been inflicted, the adjoining surface structures are freed of hair and the parts cleansed in the usual manner, (which in wounds recently inflicted, should be done without employing quantities of water) and after painting the wound surface with tincture of iodin and saturating its depths with the same agent, the wound is cleansed, if it contains filth, by means of a small curette. By using a small and sharp curette, one is enabled to cleanse the average wound quickly and almost painlessly.

In such cases, equal parts of tincture of iodin and glycerin are employed. The wound is filled with this preparation and a quantity of it is poured upon a suitable piece of aseptic gauze or cotton and this is contacted with the wound. The extremity is carefully bandaged and this dressing is left in position for forty-eight hours unless there occurs, in the meanwhile, evidence of profuse suppuration — which is unusual.

One is to be guided as to the progress made by the degree of lameness present. If little or no lameness develops, it is reasonable to expect that infection has been checked; that the wound is dry and redressing every second day is sufficiently frequent.

Where cases progress favorably, recovery (unless infectious arthritis results) should occur in from ten days to three weeks. Where extensive sub-coronary fistulae result, either from lack of prompt or

proper attention, the condition is then one requiring a radical operation to establish drainage and to disinfect if possible, the suppurating tissues.

Corns.

Etiology and Occurrence.—In horses, because of a tendency toward contraction of the heel in some subjects, together with work on hard roads and pavements, where the feet become dry and brittle, and because of neglect of the matter of shoeing, this affection is of frequent occurrence. Unshod horses are rarely affected. If conformation is faulty and too much weight is borne on the inner or the outer quarter, and the hoof wall at the quarter tends to turn inward, corns are usually present. They occur more frequently on the inner quarters of the front feet, though the outer quarters are occasionally also affected and in rare instances corns are found at the toes. They do not often affect the hind feet.

As soon as injury by pressure, such as is supposed to cause the formation of corns, is brought to bear on the sensitive sole, an extravasation of blood occurs. In time when the cause remains active, this discoloration is evident in the substance of the insensitive sole and consists in a red or yellowish spot which varies in size—this is ordinarily termed dry corn.

In some cases where infection of this extravasation of blood and serum occurs, instead of desiccation and discoloration of the insensitive parts, there is, in time, manifested a circumscribed area of destruction of the insensitive sole and the abscess may, where no provision for drainage exists, burrow between sensitive and insensitive laminae and perforate the tissues at the coronet. If the suppurative material discharges readily by way of the sole, no disturbance of the heel or quarters occurs above the hoof.

Symptomatology.—A supporting-leg-lameness characterizes this condition; and this lameness in most instances varies in degree with the amount of distress which is occasioned by pressure upon the inflamed parts. By an examination of the sole after having removed all dirt, and exposed the horny sole to view, no difficulty is encountered in locating the cause of the trouble.

Treatment.—Before suppuration has taken place and in the cases where suppuration does not occur, the horse-shoer's method of paring out the diseased tissue affords a means of temporary relief;

but unless frequently done, in many cases, lameness results within about three weeks after such treatment has been given. In other instances temporary relief is not to be gotten in this manner for any great length of time or until a more rational mode of treatment becomes necessary so that the subject may experience a cessation of the inconvenience or distress.

The general plan which meets with the approval of most practitioners consists in careful leveling of the foot and removing enough of the wall and sole at the quarters to make possible frog pressure by means of a bar shoe. With frog pressure, expansion of the heel follows in time, and permanent relief is obtainable in this manner. Thinning the wall of the quarter is advocated by many practitioners and is undoubtedly beneficial in chronic cases where marked contraction has taken place. The wall must be thinned with a rasp until it is readily flexible by compressing with the thumbs.

There are instances, however, where corns and contraction of the heel have existed so long that they do not yield to treatment. Such cases are found in old light-harness or saddle-horses that have been more or less lame for years and where there exists marked contraction of the heels, rough hoof walls, and hard and atrophied frogs.

Suppurating corns require surgical attention in the way of removal of the purulent necrotic mass and making provision for drainage. Dry dressings, such as equal parts of zinc sulphate and boric acid, may be employed to pack the cavity. After the infectious condition has been controlled, and the wound is dry, the same plan of treatment is indicated that is employed in the non-suppurating corn. Ample time is allowed, however, for the surgically invaded tissues to granulate and, if the subject is to be put in service, a leather pad, under which there has been packed oakum and tar, affords good protection.

Quittor.

This name is employed to designate an infectious inflammation of the lateral cartilage and adjoining structures. The disease is characterized by a slowly progressive necrosis and by a destruction of more or less of the cartilage and by the presence of fistulous tracts.

Etiology and Occurrence. — The disease is due to the introduction of pus producing organisms into the subcoronary region of the foot under conditions which favor the retention of such contagium and extension of infection into contiguous tissues.

Morbific material is introduced into the region of the lateral cartilage by means of calk wounds and other penetrant injuries of the foot. A sub-coronary abscess which, because of lack of proper care or because of virulency of the contagium or low vitality of the subject, is quite apt to result in cartilaginous affection and its perforation by necrosis follows.

Symptomatology. — Quittor is readily diagnosed on sight in many instances. Where there is dependable history or other evidence of the chronicity of an infectious inflammation of the kind, quittor is easily identified. If no positive evidence of the disease exists, by means of careful exploration of sinuses with the probe, one may distinguish between true cartilaginous quittor and superficial abscess formation that is often accompanied by hyperplasia.

Lameness depends upon the extent of the involvement as it affects the structures contiguous to the cartilage. A variable degree of lameness is manifested in different cases.

Treatment. — Two general plans of handling this disease are in vogue. One, the more popular method, consists in the injection of caustic solutions of various kinds into the fistulous openings with the object of causing sloughing of necrotic tissue and the stimulation of healthy granulation of such wounds. The other mode consists in either complete surgical removal of the cartilage or its remaining portions, or removal of the diseased parts of curettage.

When quittor has not extensively damaged the foot and the lateral cartilage is not partly ossified as it is in some old chronic cases, the complete removal of the lateral cartilage by means of the Bayer operation or a modification thereof is indicated. A complete description of the Bayer operation as well as Merillat's operation for this disease (the latter consisting in part, in the removal of diseased cartilage with the curette) are given in Volume three of Merillat's "Veterinary Surgical Operations."

Treatment by injection of caustic solutions has many advocates and because of the fact that, in many instances the condition is such that they are not desirable surgical cases and also because some animals may be put in service before treatment is completed, the injection method is popular.

The mode of treatment advocated by Joseph Hughes, M.R.C.V.S., constitutes a very successful manner of handling quittor and we can do no better than quote Dr. J.T. Seeley [33] on his manner of using this particular treatment.

Fig. 37—Hyperplasia of right fore foot, due to chronic quittor.

Preparation.—First remove the shoe, have the foot pared very thin and balanced as nicely as possible. Moreover, all loose fragments of horn must be detached and all crevices cleaned thoroughly.

Next, have the leg brushed and hair clipped from the knee or hock to the foot and scrubbed with ethereal soap and warm water, after which the foot must be scrubbed in like manner. The foot is then placed in a bichlorid bath several hours daily, for from two to five days, depending upon whether or not soreness is shown. The bichlorid solution is 1 to 1,000 strength.

On removing the horse from the bath a liberal layer of gauze is soaked in 1 to 1,000 bichlorid solution and placed so as to cover the entire foot. On discontinuing the bath, cover the foot with gauze saturated with a 1 to 1,000 bichlorid solution. This is to be covered with absorbent cotton and a gauze bandage, and over all is placed an oil cloth or silk covering. This pack is kept moist with bichloride solution for forty-eight hours. The foot is then ready for injection.

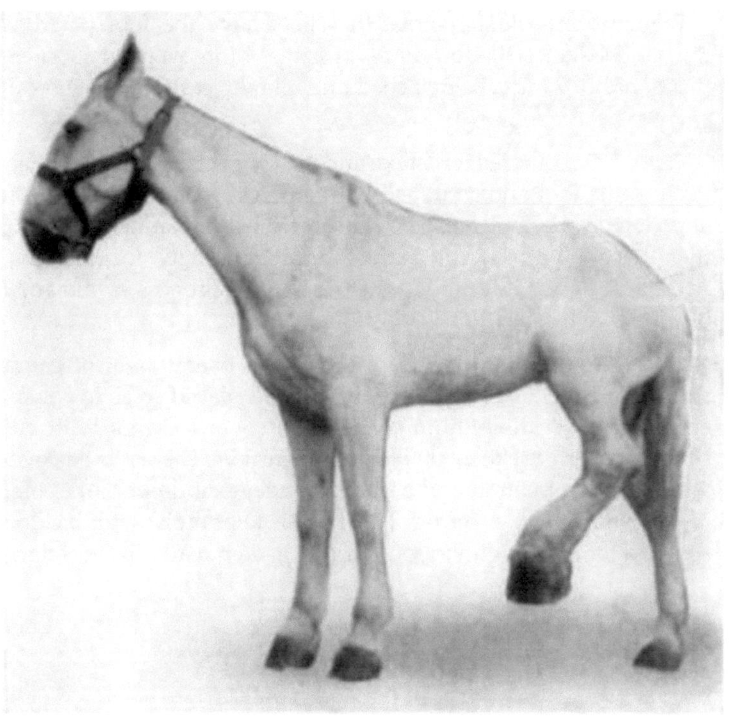

Fig. 38—Chronic quittor, left hind foot. Showing position assumed because of painfulness of the affection.

Preparation of the Injection Fluids.—Have on hand a pint of a one per cent aqueous solution of formaldehyd made under cleanly conditions, even to a clean bottle and cork, and a clean container when ready to use the liquid. Prepare also a bichlorid of mercury solution as follows: Hydrarg. Chlor. Corros. 3IV; Acid Hydrochlor. 3Iss.; Aqua Bulliens, Oij. This should be thoroughly triturated, and then filtered into a clean bottle, when it is ready for use.

Injection.—The patient should be laid on a table, if one is available, or cast, and the foot securely fixed. Then, with an ordinary one-ounce hard rubber syringe, with a good plunger (tried first to note whether or not any fluid works around between the barrel and the plunger), introduce one syringe full of the formaldehyd solution, then thoroughly probe the quittor to determine the number of sinuses. This done, inject each sinus. If two sinuses open on the surface, close one with cotton while filling the other so that if there is a connection the solution will come in contact with all tissues involved. Irrigate with the full pint of formaldehyd solution first, then follow with six or eight ounces of the bichlorid solution. Never probe the foot nor allow it to be tampered with except in the manner prescribed.

After-Treatment.—Put on a pack saturated with a solution of bichlorid of mercury 1 to 1,000 and let it remain two days. Remove pack, and once daily afterwards wipe off with cotton the secretion which accumulates on the outside, and apply a dry dressing or healing oil composed of phenol, camphor gum and olive oil.

When Dangerous to Inject.—Never inject a quittor in the acute stage. Never inject a quittor if considerable lameness is present. On injecting a solution of formalin, hold cotton tightly around the nozzle of the syringe, when the plunger is down, then withdraw the syringe gently and note particularly if the fluid returns through the opening; if none returns cease operations at once, as it is dangerous to proceed farther, it indicates that the sinus is not well defined and the fluid retained will cause much trouble and often the death of the patient.

Experience has taught that, if extensive destructive changes of the foot exist, the Bayer operation is not indicated. In the country, where quittors are not so frequently met as in urban practice, the Merillat operation is preferable in all cases. However, the cost of the protracted period of idleness, which convalescent surgical patients require, renders the Hughes method more satisfactory in the hands of the general practitioner, especially in the city.

Nail Punctures.

Nail punctures, as herein considered, embrace all penetrant wounds of the solar surface of the horse's foot due to trampling upon street nails. This does not include accidental nail pricks occasioned in shoeing. In city practice, in some stables, these cases are of frequent occurrence; and, generally speaking, nail punctures are observed more frequently in urban horses than in animals that are kept in the country.

Occurrence and Method of Examination.—This condition, then, is a rather common cause of lameness and in no case, where cause of the claudication is not obvious, is the practitioner warranted in concluding his examination without careful search for the possible existence of nail puncture of the solar surface of the foot.

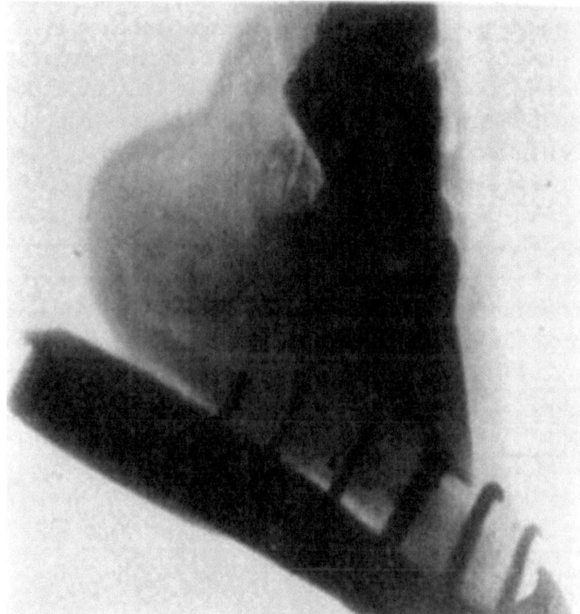

Fig. 39—Skiagraph of foot. The X-ray offers very limited possibilities in the diagnosis of lameness. The location of a "gravel" or a nail that had worked its way some distance from the surface, or of an

abscess of some proportion, deep in the tissues, might be facilitated under some circumstances by the aid of the X-ray. Its use in the detention of fractures is very limited, owing to the difficulty encountered in getting a view from the right position—many trials being necessary in most cases. The case shown above was diagnosed clinically as incipient ringbone. The X-ray revealed no lesions. (Photo by L. Griessmann.)

In occasional instances there co-exists an obvious cause for supporting-leg-lameness and an occult cause—a nail puncture. Where such complications are met, the practitioner is not necessarily guilty of neglect or carelessness when the nail puncture is not discovered at once, nevertheless, an examination is not complete until practically every possible cause of lameness has been located or excluded in any given case.

In a search for nail puncture it is necessary to expose to view every portion of the sole and frog in such manner that the existence of the smallest possible wound will be revealed. This necessitates removal of the shoe, if, after a preliminary examination, a puncture is not found, when there is good reason to suspect its presence. However, where it is readily possible to locate and care for a wound without removal of the shoe, allowing the shoe to remain materially facilitates retaining dressings in position and relieves the solar surface of contact with the ground. If extensive injury or infection exists, it is of course necessary to remove the shoe and leave it off. By removing a superficial portion of all of the sole and frog, thus carefully and completely exposing to view all parts of the solar surface of the foot, and with the aid of hoof-testers one is enabled to positively determine the existence of nail punctures. Because of the tendency of puncture wounds of the foot to close, and since the superficial portion of the solar structures are usually soiled, it is absolutely necessary to conduct examinations of this kind in a thorough manner.

Symtomatology.—Not all cases of nail puncture cause lameness during the course of the disturbance and in many instances no lameness is manifested for some time after the injury has been inflicted—not until infection has been the means of causing considerable inflammation of sensitive structures. Nevertheless, this lack of manifestation occurs only in cases where serious injury has not taken place and the degree of lameness is a constant and reliable indicator of the character and extent of nail punctures within twenty-four hours after injury has been inflicted.

The position assumed by the affected animal inconstantly varies with the location and nature of the injury and is not of particular importance in establishing a diagnosis. The subject may support some weight with the affected member and stand "base-wide" or "base-narrow," or no weight may be borne with the foot or the animal may point or keep the extremity in a state of volar flexion. In cases where extensive injury has been inflicted, and great pain exists, the foot is kept off the ground much of the time and it may be swung back and forth as in all painful affections of the extremity.

Nail punctures cause typical supporting-leg-lameness and in some cases certain peculiarities of locomotory impediment are worthy of notice. Punctures of the region of the heel, which directly affect or involve the deep tendon sheath, cause a type of lameness wherein pain is augmented, when dorsal flexion of the extremity occurs as well as when weight is borne. Wounds in the region of the toe of the hind feet sometimes cause the subject to carry the extremity considerably in advance of the point where it is planted and, just before placing the foot on the ground, it is carried backward a little way — ten or twelve inches.

However, diagnosis of nail puncture is based on the finding of the characteristic wound or resultant local changes.

Course and Prognosis. — The nature of the progress and the manner of termination of these cases are variable. If the coffin joint has been invaded, and a septic arthritis exists, the condition is at once grave. An open and infected tendon sheath, while not so serious, constitutes a condition which is distressing, and recovery is slow even under the most favorable conditions. Where a heavy, rigid and sharp nail enters the foot, in such manner that fracture of the third phalanx (os pedis) occurs, this complication makes for a protraction of the condition. Experience teaches that the natural course and termination in these cases are modified by the location and depth of the injury, virulency of the contagium and resistance of the subject to such infection.

Prevention. — In all horses which are kept at such work that exposure to nail punctures is frequent, a practical means of prevention of such injuries consists in the employment of heavy sole leather or suitable sheet metal to cover the sole of the foot and, at the same time, confine oakum and tar in contact with the solar surface to prevent the introduction of foreign material between the foot and such protecting appliances. Further, if drivers and owners could be impressed with the serious complications which so frequently attend wounds of this kind, undoubtedly many cases which are now lost, because of ignorance or neglect on the part of the teamsters or proprietors of horses, would be saved by prompt and rational treatment.

Treatment. — The treatment of this condition falls so largely within the dominion of surgery that we can give little more than an outline here.

In cases where there exists no evidence of open joint or open tendon sheath as judged by the site of the puncture and degree of lameness present (after having thoroughly cleansed the solar surface of the foot and enlarged the opening in the nonsensitive sole) a little phenol is introduced into the wound. In such cases, where it is possible for the antiseptic to contact every part of wound surface to the extreme depths of the puncture, infection is prevented when such treatment is promptly administered. This may be considered as first aid, or emergency care, and is indicated in all wounds of the foot whether the injury be serious or almost insignificant.

Subsequently one of two general courses may be pursued in the treatment of cases of nail puncture. One, by the employment of means to keep the wound patent and injection of suitable antiseptics, or agents that are more or less caustic in conjunction with strict observance of asepsis and wound protection. The other method consists in prompt establishment of drainage by surgical means and includes exploration and curettage.

The first method is better adapted to the use of the average general practitioner and he would do well to keep the opening in the nonsensitive structures patent. By introducing equal parts of tincture of iodin and glycerin daily, good results will follow in most instances. The wound is protected in unshod horses, either by completely bandaging the foot and retaining, in contact with the wound, cotton that is saturated with iodin and glycerin, or, if a minor injury exists, the moderately enlarged opening in the nonsensitive sole or frog, which has been moistened with the antiseptic, is packed with a very small quantity of cotton. A little practice in this mode of closing benign puncture wounds will enable the practitioner to successfully protect the sensitive parts in the treatment of such cases in unshod country horses.

When the condition progresses favorably the wound may be dressed every second day or twice weekly, and in the course of from two to six weeks recovery should be complete.

If the practitioner is somewhat proficient as a surgeon, and has at his command facilities for doing surgery, the second method is preferable in many cases. By using a local anesthetic on the plantar nerves and confining the subject on an operating table, restraint should be perfect. The solar surface of the foot is first thoroughly cleansed, the puncture wound is enlarged in the nonsensitive structures and the parts are then moistened with phenol or other suitable antiseptics. By means of a small probe the puncture is explored and, depending on the character of the wound and the structures involved, surgical intervention is varied to suit the case. If necessary, all of the insensitive frog is removed, and in wounds affecting the region of the heel the tissues may be incised from the puncture outward dividing all of the tissues outward and backward to the surface. A suitable surgical dressing is then applied.

If, on the other hand, the puncture extends into the navicular bursa, the radical operation is perhaps indicated, though not until one is sure that infection of the bursa and serious consequences are to follow if this operation is not performed. Detailed description of the technic of this operation belongs to the realm of surgery and a good discussion of it is to be found in William's work on veterinary surgical and obstetrical operations.

One may summarize the discussion of treatment of nail puncture by saying that emergency care as herein described is of first consideration. In every case an immunizing dose of anti-tetanic serum should be given. Subsequently, the method employed must suit the character of the wound, existing facilities for handling the subject and the skill and aptitude of the practitioner.

FOOTNOTES:

[5] Manual of Veterinary Physiology, by Major-General F. Smith, page 590.

[6] Manual of Veterinary Physiology by Major-General F. Smith, page 589.

[7] Regional Veterinary Surgery and Operative Technique, Jno. A.W. Dollar, M.R.C.V.S., F.R.S.E., M.R.I., page 765.

[8] Dr. Roscoe R. Bell in the Proceedings, N.Y. State Veterinary Medical Society, 1899.

[9] American Veterinary Review, Vol. 35, P. 456.

[10] "Radial Paralysis and Its Treatment by Mechanical Fixation of Knee and Ankle," Geo. H. Berns, D.V.S. Proceedings of the American Veterinary Medical Association, 1912, p. 219.

[11] As quoted by Berns, in Radial Paralysis, etc., Proceedings of the A.V.M.A., 1912.

[12] Veterinary Surgical Operations, by L.A. Merillat, V.S., p. 507.

[13] A paper presented before the Illinois Veterinary Medical Assn. by Dr. H. Thompson of Paxton, Ill., American Veterinary Review, Vol. 15, p. 134.

[14] "Fractures in Foals," by Dr. Wilfred Walters, M.R.C.V.S., American Journal of Veterinary Medicine, Vol. 8, p. 669.

[15] American Veterinary Review, Vol. 26, p. 1068.

[16] Fractures, by H. Thompson, Paxton, Ill., American Veterinary Review, Vol. 15, p. 134.

[17] Veterinary Surgical Operations, by L.A. Merillat, Vol. 3, p. 198.

[18] Wilfred Walters, American Journal of Veterinary Medicine, Vol. 8, p. 606.

[19] J.N. Frost, assistant professor of Surgery, Veterinary Dept., Cornell University, in "Wound Treatment," page 159.

[20] Open Joints and Their Treatment in my practice, by J.V. Lacroix, American Journal of Veterinary Medicine, Vol. 5, page 203.

[21] Regional Veterinary Surgery Möller—Dollar, page 605.

[22] Extract from Receuil de Médecine Vétérinaire in Ameircan Veterinary Review, Vol. 23, p. 893.

[23] Fracture of All the Sesamoid Bones, by R.F. Frost, M.R.C.V.S., A.V.D., Rangoon, Burmah, in American Veterinary Review, Vol. 5, p. 362.

[24] The Anatomy of the Domestic Animal, by Septimus Sisson, S.B., V.S.

[25] Traité De Thérapeutique Chirurgicale Des Animaux Domestique, par P.J. Cadiot et J. Almy, Tome Second, page 547.

[26] Anatomie Regionale Des Animaux Domestique, page 695.

[27] Manual of Veterinary Physiology, by Major-General F. Smith, C.B., C.M.G., page 678.

[28] Möller's Regional Veterinary Surgery, by Dollar, page 630.

[29] Edinburgh Veterinary Review, Vol. VI, page 616.

[30] Equine Laminitis or Pododermatitis, by R.C. Moore, D.V.S., American Journal of Veterinary Medicine, Vol. XI, page 284.

[31] American Journal of Veterinary Medicine, Vol. XI, page 318.

[32] The Shoeing of a Dropped Sole Foot by Dr. David W. Cochran, New York City, The Horse Shoers Journal, March, 1915.

[33] Quittor and Its Treatment by the Hughes Method, J.T. Seeley, M.D.C., Seattle, Washington, Chicago Veterinary College Quarterly Bulletin, Vol. 9, page 27.

SECTION IV.

LAMENESS IN THE HIND LEG.

Anatomo-Physiological Consideration of the Pelvic Limbs.

The pelvic bones as a whole constitute the analogue of the scapulae with respect to their function as a part of the mechanism of locomotive and supportive apparatus of the horse. The manner of attachment or connection between the ilia and the trunk is materially different from that of the scapulae, however, and the angles as formed by the long axes of the ilia in relation to the spinal column are maintained by two functionally antagonistic structures—the sacrosciatic ligaments, and the abdominal muscles by means of the prepubian tendon. The sacro-iliac articulations are such that a very limited amount of movement is possible; free movement, however, is unnecessary because of the enarthrodial (ball and socket) femeropelvic joint.

The various muscles which exert their effect upon the pelvis in changing their relationship between the long axes of the ilia and spinal column, are concerned but little more in propulsion and weight bearing than are the pectoral muscles. A general treatise on the subject of lameness does not properly include such structures any more than it does the various affections of the dorsal, lumbar and sacral vertebrae or inflammation of the abdominal parietes. Involvement of such parts cause manifestations of lameness but the matter of establishing a diagnosis is difficult in many instances and in some cases impossible.

The femeropelvic articulation is formed by the hemispherical head of the femur and the acetabulum; the latter constituting a cotyloid cavity which is deepened by the cotyloid ligament.

The round ligament (ligamentum teres) is the principal binding structure of the hip joint and it arises in a notch in the head of the femur and is attached in the subpubic groove close to the acetabular notch. Another ligament, peculiar to Equidae—the accessory (pubiofemoral)—is attached to the head of the femur near the round ligament and passes through the cotyloid notch and along the under side of the pubis. It is inserted or blends with the prepubic tendon. This ligament prevents extreme abduction of the leg. The joint capsule encompasses the articulation and is attached to the brim of the acetabulum and the edge of the head of the femur.

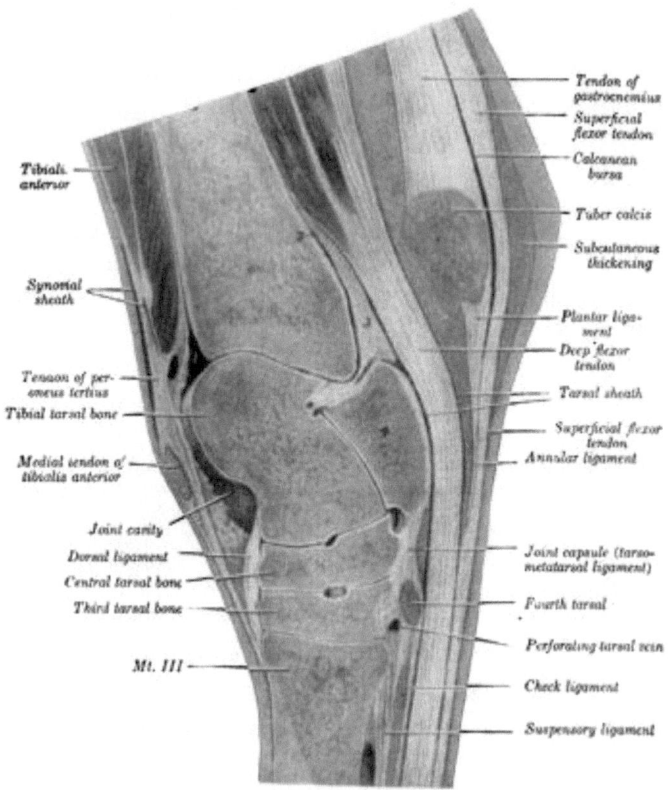

Fig. 40—Sagital section of right hock. The section passes through the middle of the groove of the trochlea of the tibial tarsal bone. 1 and 2. Proximal ends of cavity of hock joint. 3. Thick part of joint capsule over which deep flexor tendon plays. 4. Fibular tarsal bone (sustentaculum). A large vein crosses the upper part of the joint capsule (in front of 1). (From Sisson's "Anatomy of the Domestic Animals.")

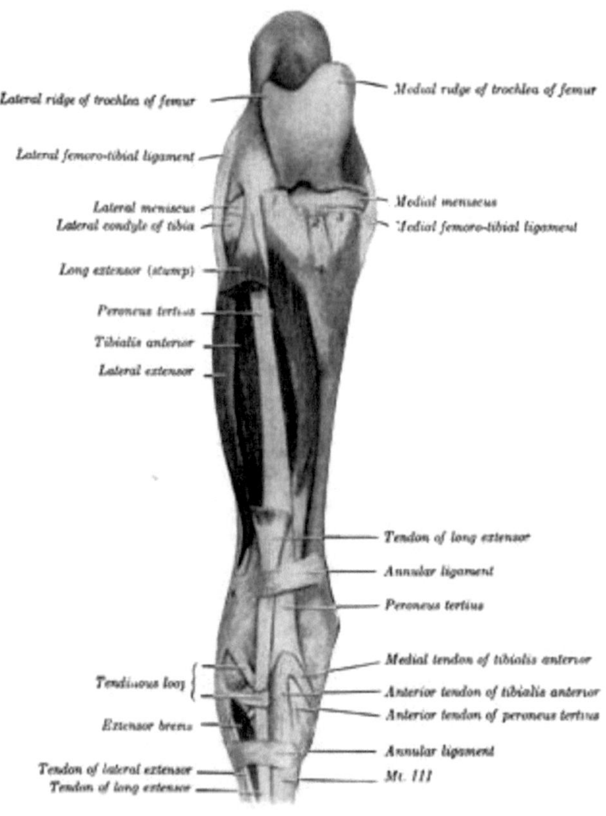

Fig. 41—Muscles of right leg; front view. The greater part of the long extensor has been removed. 1, 2, 3. Stumps of patellar ligaments. 4. Tuberosity of tibia. (From Sisson's "Anatomy of the Domestic Animals.")

The stifle joint is analagous to the knee joint of man and is to be considered an atypical ginglymus (hinge) articulation formed by the femur, tibia and patella. The ligaments are femerotibial, femeropatellar and capsular.

In addition to the usual provision for articulation of bones there are situated cartilaginous *menisci* between the condyles of the femur and the head of the tibia. These discs surround the tibial spine and are otherwise shaped to fit perfectly between the articular portions of the femur and tibia.

Collateral ligaments (internal and external lateral) pass from the distal end of the femur to the proximal portion of the tibia. The mesial (internal) arises from the internal condyle of the femur and is attached to a rough area below the margin of the medial (internal) condyle of the tibia. The lateral (external), shorter and thicker, arises from the depression on the lateral epicondyle and inserts to the head of the fibula.

The crucial or interosseus, anterior and posterior, are situated between the femur and tibia, and according to Smith,[34] the crucial ligaments are necessary to properly join the two bones, because of the character of the structure of the articular ends of the femur and tibia.

The femeropatella ligaments are two thin bands which reinforce the capsular ligament. They arise from the lateral aspects of the femur, just above the condyles and are inserted to the corresponding surfaces of the patella.

The patellar ligaments are three strong bands which arise from the antero-inferior surface of the patella, and are inserted to the anterior aspect of the tuberosity of the tibia.

Taken as a whole, the tarsal bones, interarticulating and articulating with the tibia and metatarsal bones form the hock joint and this articulation is analagous to the carpus. As with the carpus, there is less movement in the inferior portion of the joint than in the superior part of the articulation. The chief articulating parts are the tibia with the tibial tarsal bone (astragulus).

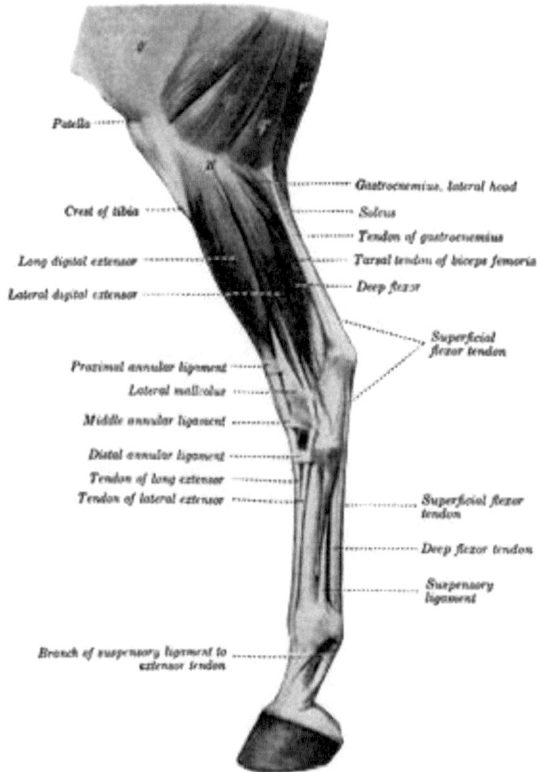

Fig. 42—Muscles of lower part of thigh, leg and foot; lateral view, o', Fascia lata; q, q', q", biceps femoris; r, semitendinosus; 21', lateral condyle of tibia. The extensor brevis is visible in the angle between the long and lateral extensor tendons. (After Ellenberger-Baum, Anat. für Künstler.) (From Sisson's "Anatomy of the Domestic Animals.")

The capsular ligament is attached around the margin of the articular surfaces of the tibia, to the tarsal bones, the collateral ligaments (internal and external lateral) and to the metatarsus.

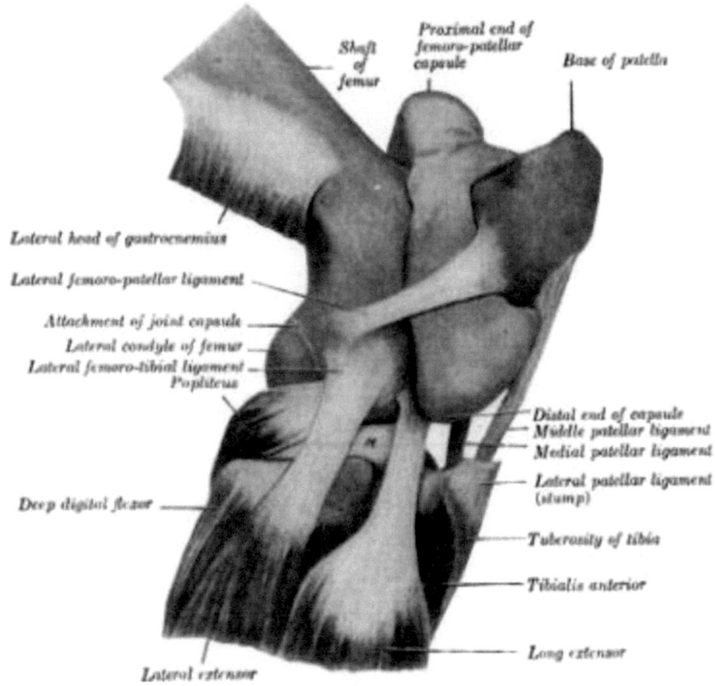

Fig. 43 — Right stifle joint; lateral view. The femoro-patellar capsule was filled with plaster-of-Paris and then removed after the cast was set. The femoro-tibial capsule and most of the lateral patellar ligament are removed. M. Lateral meniscus. (From Sisson's "Anatomy of the Domestic Animals.")

The common ligaments of the tarsal joint are the collateral, the plantar (calcaneo-metatarsal and c. cuboid) and dorsal ligaments (oblique).

The medial (internal lateral) ligament serves to join the medial (internal) tibial malleolus with tibial tarsal (astragalus) and other tarsal bones.

The lateral (external lateral) ligament is inserted to the lateral (external) tibial malleolus and its distal portions are attached to the tibial tarsal (astragalus), fibular tarsal (calcaneum) bone, fourth tarsal (cuboid) and metatarsus bones.

262

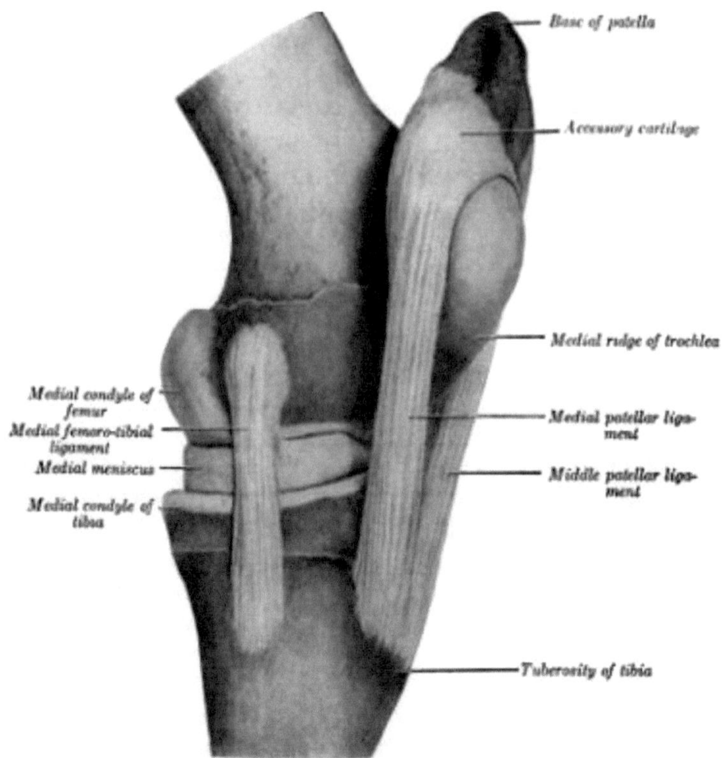

Fig. 44—Left stifle joint; medial view. The capsules are removed. (From Sisson's "Anatomy of the Domestic Animals.")

The plantar ligament (calcaneo-cuboid) is a strong flat band which is attached to the plantar surface of the fibular and fourth tarsal bones (calcaneum and cuboid) and the head of the lateral metatarsal (external small) bone.

The dorsal (oblique) ligament is attached above to the distal tuberosity on the inner side of the tibia. It is inserted below to the central (cuneiform magnum) and third (c. medium) tarsal bones, to the proximal ends of the large and outer small metatarsal bones.

The tarsus is a true hinge joint and because of the great strain which it sustains, is subject to frequent injury. About seventy-five percent of cases of lameness affecting the hind leg may be said to arise from disease of the hock.

As members of locomotion the legs receive strains of two kinds: those of concussion and weight-bearing and strains of propulsion; the latter are the greater. In the horse as a work animal, the hind legs are probably subjected to greater strains than are the front but the manner of construction of the various parts of the pelvic limbs with the possible exception (according to some authorities) of the tibial tarsal joint, offsets this condition.

The femur may be considered analagous to the humerus in that it bears a similar relationship to the ilium, that exist between the humerus and scapula. Further flexion during repose is prevented chiefly by the glutens medius (maximus) muscle and its tendons. The larger tendon inserts to the summit of the trochanter major of the femur and corresponds to the biceps brachii in the action of the latter on the scapulohumeral joint, except that the gluteus medius, in attaching to the femoral trochanter, exerts its effect as a lever of the first class. Because of the relationship between the long axes of the femur and iliac shaft it is evident that the angle formed by these two bones is maintained chiefly by the gluteus muscles during weight bearing. Contraction of muscular fibers of the gluteus medius causes extension of the femur and muscular strain is prevented to a great degree by the inelastic portion of this muscle. The chief physiological antagonistics of the glutei are the quadriceps femoris and tensor fascia lata.

While the leg is supporting weight the stifle joint is fixed in position mainly by the quadriceps femoris group of muscles which are attached to the patella. Tendinous fibres intersect this muscular mass and relieve muscular strain during weight bearing. Because of the manner in which the patella functionates with the trochlea of the femur, comparatively little energy is required to prevent further flexion of the stifle joint. The patella, according to Strangeways, may be considered a sesamoid bone.

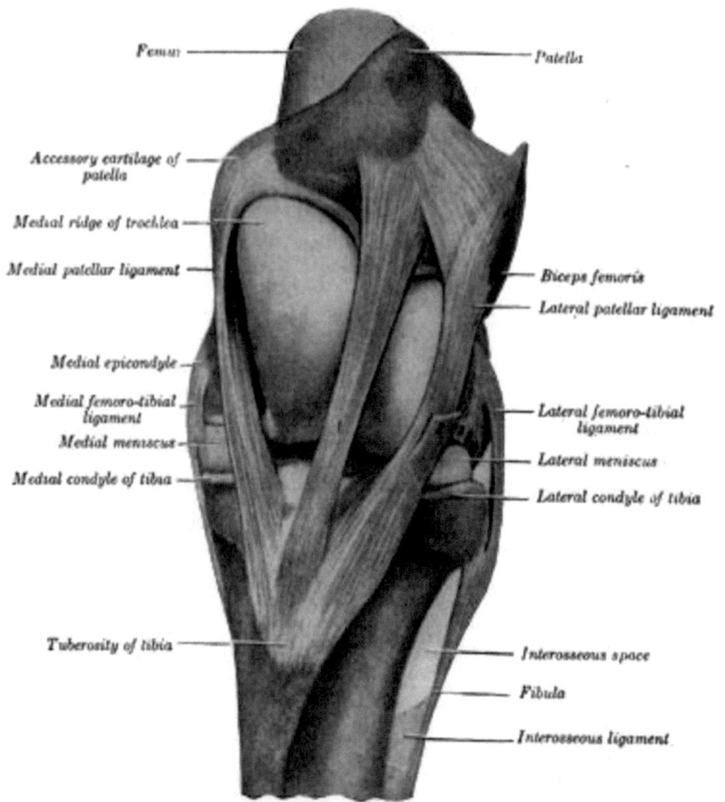

Fig. 45—Left stifle joint; front view. The capsules are removed. 1. Middle patellar ligament. 2. Stump of fascia lata. 3. Stump of common tendon of extensor longus and peroneus tertius. (From Sisson's "Anatomy of Domestic Animals.")

The quadriceps group of muscles is assisted by the anterior digital extensor (extensor pedis) peroneus tertius and tibialis anticus (flexor metatarsi) muscles. The latter pair (flexor metatarsi, muscular and tendinous portions, because of their attachment to the external condyle of the femur and to the metatarsal bone) are enabled to automatically flex the tarsal joint when the stifle is flexed.

The hock is kept fixed in position by the gastrocnemius and the superficial digital flexor (perforatus). The latter structure, which is chiefly tendinous, originates in the supracondyloid fossa of the femur and has an insertion to the summit of the fibular tarsal (calcis) bone. It relieves the gastrocnemius of muscular strain during weight bearing.

Smith[35] styles the function of the stifle and hock joints a reciprocating action, and we quote from this authority the following:

From what has been said, it is evident that flexion and extension of stifle and hock are identical in their action. When the stifle is extended, the hock is automatically extended, nor can it under any circumstances flex without the previous flexion of the stifle. There is no parallel to this in the body. The two joints, though far apart, act as one, and they are locked by the drawing up of the patella, and in no other way. The so-called dislocation of the stifle in the horse is a misnomer. That the patella is capable of being dislocated is beyond doubt, but the ordinary condition described under that term, when the stifle and hock are rigid while the foot is turned back with its wall on the ground, is nothing more than spasm of the muscles which keeps the patella drawn up. The moment they relax the previously immovable limb and useless foot have their function restored as if by magic, but are immediately thrown out of gear in the course of a few minutes as a recurrence of the tetanus of the petallar muscle takes place. The fascia of the thigh, like that of the arm, is a most potent factor in giving assistance to the constant strain imposed on the muscles of the limbs during standing.

Below the hock the hind limb is arranged like that of the fore, the deep flexor (perforans) receiving its additional support from the "check ligament," as in the fore leg.

The natural attitude of standing adopted by the horse is to rest on three legs—one hind and two fore. If he is alert, he stands on all four limbs; but if standing in the ordinary manner, he always rests on one hind leg. He does not remain long in this position without changing to the other. Hour by hour he stands, shifting his weight at intervals from one to the other hind leg, and resting its fellow by flexing the hock and standing on the toe. He never spares his forelimbs in this manner in a state of health, but always stands squarely on them.

Hip Lameness.

Fortunately, because of the heavy musculature which goes to form a part of the locomotive apparatus of the rear extremity, hip lameness is comparatively rare. While the term is in itself ambiguous and signifies nothing more definite than does "shoulder lameness," yet diagnosis of almost any condition that may be classed under the head of "hip lameness" is not easy except in cases where the cause is obvious, as in wounds of the musculature and certain fractures. To the complexity which the gait of the quadruped contributes, because of its being four-legged, there is added the complicated manner of articulation of the bones of the hind leg. This involves the hip in the manner of diagnostic problems and because of the inaccessibility of certain parts, owing to the bulk of the musculature of these parts, diagnosis of some hip ailments becomes an intricate problem. Consequently, in some instances, before one may arrive at definite and enlightening conclusions, repeated examinations are necessary as well as a knowledge of reliable history and recorded observations of the subject over a considerable period.

Rheumatic affections, when present, usually cause recurrent attacks of lameness; myalgia, due to subsurface injury occasioned by contusion, generally produces an ephemeral disturbance; and while these are examples of cases where occult causes are active, they are by no means unprecedented. In cases where the cause of lameness is not definitely located, and when by the process of exclusion one is enabled to decide that the seat of trouble is in the hip, a tentative diagnosis of hip lameness is always appropriate.

In one instance a Shetland pony evinced a peculiar form of intermittent lameness which affected the left hip, and repeated examinations did not disclose the cause of the trouble. After about a year there was established spontaneously an opening through the integument overlying the region of the attachment of the psoas major (magnus), through which pus discharged. With the occurrence of this fistula, lameness almost entirely disappeared, but the emission of a small amount of pus persisted for more than a year. The subject was not observed thereafter and the outcome in this case is not a matter of record. Whether there existed a psoic phlegmon due to metastatic infection or necrosis of a part of a lumber or dorsal vertebra is a matter for speculation. Thus the presence of some anomalous conditions which affect the pelvic region and cause lameness may be discovered, yet both in hip and shoulder regions causes may not be definitely located by means of practical methods of examination.

Injuries of all kinds are the more frequent causes of hip lameness. In such cases, lameness may result directly and resolution be prompt, or the claudication become aggravated in time, due to muscular atrophy or degenerative changes affecting the hip joint or nerves. Rheumatism or metastatic infection may be the cause of hip lameness as well as affections of the pelvic bones, lumbar and sacral vertebrae. Hip lameness may also be provoked by melanotic or other tumors.

In the diagnosis of hip lameness, one is guided in a general way by the character of the impediment manifested. Swinging-leg lameness is often present and the impediment is more accentuated when the animal is caused to step backward. In many cases lameness is mixed, being about equally noticeable during weight bearing and while the member is being swung. By exclusion of causes which might affect other parts; one may definitely locate the cause of the trouble or determine that a certain region is affected.

The sudden manifestation of lameness is indicative of injury; thermic disturbances may signalize metastatic infection; history, if dependable, is always helpful. Repeated observations, taking into account the course which the affection assumes during a period of a few days, often serve to afford a means of establishing a diagnosis in baffling cases.

Fractures of the Pelvic Bones.

The os innominatum may be so fractured that the pelvic girdle is broken, as in fracture of the iliac shaft, or in a manner that the girdling continuity of the innominate bones is not interrupted. It naturally follows that greater injury is done when the pelvic girdle is broken than when it is not, except in cases where the acetabulum is involved and its brim not completely divided.

Etiology and Occurrence.—Pelvic fractures are usually caused by falls or other manner of contusion. Cases are reported where it would seem that fracture of the iliac angle resulted from muscular contraction, but it is certain that most fractures of this kind are due to collisions with door jambs or similar injuries. In old horses especially, fracture of pelvic bones occurs frequently. This form of injury is of more frequent occurrence in animals of all ages that work on paved streets. The country horse is not subjected to the uncertain footing of the slippery pavement, nor to injuries which compare with those caused by contusions sustained in falling upon asphalt or cobble-stones.

Symptomatology.—While in many cases of pelvic fracture lameness or abnormal decumbency are the salient manifestations, yet the pathognomic symptoms are crepitation or palpable evidence which may be obtained by rectal or vaginal examination. In fractures of the angle of the ilium and the ischial tuberosity, perceptible evidence always exists.

In cases where fracture of some portion of the pelvic girdle is suspected and the subject is able to walk, crepitation is sought by placing one hand on an external angle of the ilium and the other on the ischial tuberosity and the animal is then made to walk. Or, by placing the hands as just directed, an assistant may grasp the horse's tail and by alternately exerting traction on the tail and pushing against the hip in such manner that weight is shifted from one leg to the other, crepitation may be detected.

Fracture of the pubis near its symphysis constitutes a grave injury, as there is danger of the bladder becoming caught in the fissure and perforation of its wall may result. Such a case is reported by

Bauman[36] wherein a three-year-old gelding bore the history of having been lame for ten days. Upon rectal examination the bladder was found to be hard and tumor-like and about the size of a baseball. The body of the ischium in this case was fractured and a rent in the bladder was caused by a sharp projecting piece of bone. Autopsy revealed, in addition to the fracture and rent of the bladder wall, a large quantity of urine in the peritoneal cavity.

In other instances hemorrhage caused death and not infrequently infection was responsible for a fatal issue. Moller,[37] quoting Nocard, describes a case where fracture occurred through the region of the foramen ovale and paralysis of the obturator nerve followed.

Fractures which include the acetabular bones cause great pain. This is manifested by marked lameness, both during weight bearing and when the member is swung. Such cases terminate unfavorably—complete recovery is impossible.

Where small portions of the angle of the ilium are broken, and the skin is left intact, there exists the least troublesome class of pelvic fracture. If large portions of the ilium are fractured, considerable disturbance results. There eventually occurs more or less displacement in such cases, if such displacement does not take place at the time of injury. The same may be said of fracture of the tuber ischii, but when these bones are fractured a more serious condition results.

Treatment.—When a case is found to be uncomplicated, that is, if the fracture is such that recovery seems possible and after having determined that treatment may be practicable, the first consideration is that of confining the subject in suitable slings. In many cases of pelvic fracture, the affected animal will need to be kept in slings from six weeks to three months, and it becomes a difficult problem to minimize the distress during this long period of confinement in the peculiar manner required for favorable outcome.

The pattern of sling employed should be the best that is obtainable and the matter of its adjustment is quite important lest unnecessary chafing or even necrosis of skin result. Frequent readjustment may be necessary, and time is well spent in this manner since this contributes materially toward a favorable termination by encouraging the subject to remain quiet so that coaptation of the broken bones may be maintained. Aside from slings, mechanical appliances that are helpful in the treatment of these cases are not yet in use.

A regimen that is nutritive and at the same time laxative is essential and in some cases cathartics and enemata are necessary. Also, during the first few days, if there is retention of urine, catheterization is imperative. In a word, the handling of such cases consists largely in keeping the subject inactive, as comfortable as possible, and giving attention to suitable diet.

Simple fracture of the external iliac angle needs no particular attention, except that the subject is kept quiet until lameness subsides. In all cases where much of the bone is broken, the animal is blemished, but interference with function does not follow. If infection results because of a compound fracture, loose pieces of bone must be removed surgically and drainage provided for.

In fracture of the ischial tuberosity, infection is more apt to result than in like injury of the ilium, and greater displacement of bone occurs. This displacement, due to contraction of the attached muscles, is in some instances a contributing cause to the infection which often follows in these cases. In females where the body of the ischium is fractured, lacerations of the vagina may be present, and this constitutes a serious complication which usually terminates fatally.

After-care in fracture of the pelvic girdle consists principally in allowing a protracted period of rest before subjects are put to work.

Fractures of the Femur.

Etiology and Occurrence. — This is a comparatively rare injury in the horse because of the protection afforded the femur by the heavy musculature. Fragilitas of the bone probably exists in many cases when fracture of its diaphysis occurs. It is generally conceded that the neck of the femur is rarely broken because of a lack of constriction in this part, but fracture of the trochanters has been recorded rather frequently. However, Lienaux and Zwanenpoete[38] state that fracture of the neck of the femur is of frequent occurrence in Belgian colts. Tapley[39] reports in the Veterinary Journal (English) fracture of the head and internal trochanter of the femur and patellar luxation occurring simultaneously affecting a mule. In this case the mule was found decumbent on a concrete floor. After three weeks, the subject was destroyed and autopsy revealed rupture of the left pubiofemoral ligament, tearing with it a portion of the articular surface of the femur. The internal trochanter was also fractured in four small pieces. In this case it is fair to suppose that the mule in trying to regain footing on a slippery floor violently abducted the legs and fracture resulted. It is possible also that a temporary luxation of the patella took place first and caused the animal to struggle in such manner that fracture followed.

Fig. 46 — Oblique fracture of the femur of a 1,500 six-year-old draft horse. Showing shortening of bone, owing to a lateral approximation of the diaphysis because of muscular contraction. Photo by Dr. Edward Merillat.

Symptomatology. — According to Cadiot and Almy,[40] "regardless of the location of femoral fractures, the subject is usually intensely lame, the animal frequently walking on three legs — fractures of the diaphysis are characterized by an abnormal mobility."

As a rule, crepitation is to be recognized in fractures of the shaft of the bone, by passively moving the leg to and from the medial plane (adduction and abduction).

Fracture of the trochanter major is signalized by local swelling and evidence of pain; the forward stride is shortened because this movement tenses the tendon of the gluteus major (maximus) which is attached principally to the trochanter.

Fig. 47 — Same bone as in Fig. 46 after about six months' treatment. In this case Dr. Merillat employed a weight to counteract muscular contraction. It is noticeable that very little provisional callus has formed in this case, and in spite of unusual ingenuity and good facilities for caring for the subject, union of bone did not occur.

Treatment. — Reduction of femoral fracture in the horse is practically impossible, and retaining the broken bones in coaptation is not possible by means of mechanical appliances. Consequently, prognosis is unfavorable in fracture of the body of the femur. When union of bone occurs, there results shortening of the leg and animals are rendered permanently lame. If the immediate region of the head of the bone is involved as well as in case of fracture of the condyles, an incurable arthritis ensues.

Where the trochanters are broken, chronic lameness and muscular atrophy is the result. Therefore, it is evident that, because of the manner of function of the femur, the leverage afforded by its great trochanter and its heavy muscular attachments, fractures of this bone in the horse do not terminate favorably.

Luxation of the Femur.

Etiology and Occurrence. — Uncomplicated femoral luxation is of less frequent occurrence in the horse than in the other domestic animals. The deep cotyloid cavity renders disarticulation difficult and luxation does not often take place. Complications that usually occur are rupture of the round (coxofemoral) ligament or fracture of the neck of the femur. Falls or violent strains are necessary to produce this luxation. Goubaux is quoted by Cadiot and Almy[41] as having observed the head of the femur in an instance wherein luxation had long existed. In this case autopsy revealed the fact that the inner portion (two-thirds) of the head of the femur had completely disappeared.

Luxation of the femur is observed in old emaciated animals that are worked on slippery pavements. Occasionally, evidence of chronic luxation of the femur is observed in the anatomical laboratory. The chronicity of the condition is obvious when one notes the well formed articulation which Nature provides for the head of the femur, where fracture or other serious complications are not present.

Symptomatology. — In every case there must exist either restriction of movement or an evident abnormal position of the leg, or both conditions may exist at once. Also, the leg may be markedly shortened. Manifestation of this affection varies, depending upon the character of the luxation (position of the head of the humerus with relation to the acetabulum). Lusk[42] cites a case of a mule which had suffered femoral luxation. The animal was destroyed and on autopsy the head of the femur found to be contained within a false articular cavity situated about four inches above the acetabulum. In Dr. Lusk's case as he states it, the following symptoms were presented: "Limb shortened and fixed in a position of adduction. While standing the affected limb hung directly across and in front of the opposite one; upper trochanter very prominent; skin over hip joint very tense. The mobility of the limb was very limited, especially in the forward direction."

Being very prominent when there is an upward luxation and less perceptible in downward displacement, the location of the trochanter major is an indicator of the character of the luxation with respect to the position of the head of the femur. This variation of position causes abnormal tenseness or looseness of the skin over the region of the trochanter major. Rectal examination is of aid in locating the head of the humerus.

Treatment.—When it is evident that a subject should be given treatment and not destroyed, the animal must be cast and completely anesthetized. With complete relaxation thus secured by rotation of the limb, using the hip joint region as a pivot, reduction may be effected. Traction is exerted in the same direction from the acetabulum that the head of the femur is situated and by pressing over the joint, the displaced bone may be returned in position. If luxation is downward, traction on the extremity will tend to dislodge the head of the femur from the inferior acetabular margin making reduction possible.

The same general plan which is ordinarily employed in correcting luxation is indicated here, but because of the heavy musculature of the hip, complete anesthesia is imperative in all such manipulations.

Gluteal Tendo-Synovitis.

The glutens medius (g. maximus) muscle is inserted chiefly by means of two tendons; one to the summit of the trochanter major of the femur and the other passing over the anterior part of the convexity of the trochanter, and being attached to the crest below it. The trochanter is covered with cartilage, and a bursa (the trochanteric) is interposed between the tendon and the cartilage.

Etiology and Occurrence.—This affection is probably caused in most instances by direct injury to the parts, such as may be occasioned by being kicked, falling on pavement, or being struck by the body of a heavy wagon. Strains in pulling or in slipping are undoubtedly causative factors and in draft horses such strains may result in involvement of this synovial apparatus.

Symptomatology.—If pain be severe and inflammation acute, weight may not be borne with the affected member. There is some local manifestation of the condition in acute cases. Swelling of the tissues contiguous to the bursa is present and pain is evinced upon manipulation of the parts. A characteristic gait marks inflammation of the trochanteric bursa, and as Gunther has put it, the subject generally moves or trots as does the dog—the sound member being carried in advance of the affected one and the forward stride of the diseased leg is shortened. In some chronic cases crepitation is discernible by holding the hand on the trochanter while the subject walks.

Treatment.—In the first stages of an acute affection absolute quiet must be enforced; local antiphlogistic applications are beneficial. Later, vesication of a liberal area surrounding the trochanter major is indicated. Where the condition has become chronic in horses that are to be kept at heavy draft work there is little chance for complete recovery. And, naturally, one is not to expect resolution in cases where there exist erosion and ossification of cartilage—where crepitation is discernible.

Paralysis of the Hind Leg.

Aside from paraplegic conditions due to disease of the cord or the lumbosacral plexus, and monoplegic affections resultant from disturbances of this plexus, paralysis of certain nerves are occasionally encountered.

Anatomy.—The lumbosacral plexus results substantially from the union of the ventral branches of the last three lumbar and the first two sacral nerves, but it derives a small root from the third lumbar nerve also. The anterior part of the plexus lies in front of the internal iliac artery, between the lumbar transverse processes and the psoas minor. It supplies branches to the iliopsoas[43] (designated by Girard, the iliacomuscular nerves). The posterior part lies partly upon and partly in the texture of the sacrosciatic ligament. From the plexus are derived the nerves of the pelvic limb (Sisson).

Paralysis of the Femoral (Crural) Nerve.

Anatomy.—The femoral nerve (crural) is derived chiefly from the fourth and fifth lumbar nerves. It runs ventrally and backward, at first between the psoas major and minor, then crosses the deep face of the tendon of the latter and descends under cover of the sartorious over the terminal part of the iliopsoas. It innervates the psoas major (magnus), psoas minor (parvus), sartorious, rectus femoris, vastus lateralis (interims). Branches supply the stifle and the adductor and pectineus muscles.

Etiology and Occurrence.—While paralysis of the femoral nerve, also known as "dropped stifle" occurs as a result of local injuries and melanotic tumors in gray horses, most cases are due to azoturia. So-called crural paralysis or "hip swinney" is occasionally observed but this is not a condition wherein the nerve is affected in the manner that characterizes the marked atrophy of quadriceps femoris (crural) muscles in some cases of hemaglobinuria. This form of paralysis according to Hutyra and Marek is due primarily to diffuse degeneration of the muscles.

Symptomatology.—When muscular atrophy is not extensive no particular evidence of this condition may be manifested while the subject is at rest, but where muscular waste has occurred, the nature of the ailment is at once recognized. Since the femoral nerve supplies the quadriceps femoris muscles, it follows that when the psoic portion of this nerve becomes diseased, the stifle loses its support, and in a unilateral involvement when the subject attempts to walk on the affected member, the stifle sinks down for want of support and the leg collapses unless weight is caught up with the other leg. Often, following azoturia, a bilateral affection is to be observed.

Treatment.—Horses may be restrained in the standing position, and in the average instance, a twitch and hood are all the restraining appliances necessary.

In cases where the disease is unilateral and atrophy is not of too long standing, recovery is possible in vigorous subjects. All affections, however, wherein degenerative changes involve the nerve trunk, whether due to diffuse myositis or pressure from malignant tumors, will not yield to treatment.

The same general plan of treatment is indicated that is described on page 74 in the consideration of atrophy of the scapular muscles. It is especially important to provide for the subject to be exercised when there is atrophy of the quadriceps muscles following azoturia.

In addition to the foregoing, good results have attended the use of intramuscular injections of oxygen. The technic of the operation consists in preparing the area of skin which covers the atrophied muscles as for any operation. The hair is clipped over five or six or more circular areas of about an inch in diameter; the skin is cleansed and then painted with tincture of iodin.

A long heavy sterile needle, which is connected with an oxygen tank by means of six feet of rubber tubing, is thrust into the depths of the affected muscles and the gas is gently introduced into the tissues. One needs exercise extreme care that the gas enter slowly because great pain is produced by the sudden injection of the oxygen. Likewise too much of the gas must not be introduced at one place. When the oxygen is slowly introduced it may be allowed to enter the tissues until the subject gives evidence of experiencing considerable pain, or if the parts are not particularly sensitive, a reasonable amount (enough to cause a mild degree of diffuse inflammation) is introduced at each one of five or six points. In large animals more points of injection may be used.

No infection or other bad results will follow the execution of a good technic and the treatment may be repeated every three or four weeks until either marked regeneration of tissue is evident or the case is obviously proved hopeless.

Paralysis of the Obturator Nerve.

Anatomy.—The obturator nerve, situated at first under the peritoneum, accompanies the obturator artery through the obturator foramen and gaining the muscles on the internal face of the thigh, terminates in the obturator externus, adductors, pectineus and gracilis, also giving twigs to the obturator internus (Strangeways).

Etiology and Occurrence.—This condition occurs upon rare occasions as the result of injury such as falls which cause extreme abduction of the legs, or in pelvic fracture where the nerve is directly injured, or when melanotic tumors or other new growths compress the nerve in such manner that its function is suspended. Paralysis of the obturator nerve or nerves is met with rather frequently, notwithstanding, in mares, following dystocia. The nerves (one or both) may become bruised at the brim of the obturator foramen by being caught between the pelvis and the body of the fetus in some cases of protracted labor.

Symptomatology.—In a unilateral affection there may be little evidence of the trouble while the subject is standing; or there is to be seen some abduction; or the affected member may present abduction of the stifle and stand "toe outward." If the animal is walked there will be manifested more or less abduction and the character of the impediment varies according to the nature of the involvement.

Following protracted cases of labor in some instances where only a unilateral paralysis exists, walking is performed with difficulty; the subject may be unable to support weight with the affected member and is obliged to hop on the one sound hind leg. In bilateral affections, they are unable to rise. If the condition is severe the sling is required to keep the subject standing, and with this care, recovery will follow.

Treatment.—If new growths or callosities or similar conditions affect the nerve, little, if any, hope for recovery exists. In young and vigorous subjects where cause is not definitely known, a course of strychnin may be given. Good nursing, providing for the subject's comfort and allowing moderate exercise, constitute rational treatment. Stimulating embrocations on the abductor muscles resorted to in cases during the incipient stage may prove helpful.

When paralysis of the obturator nerve occurs as a post-partum complication, and other conditions are favorable, the subject should be raised to its feet without unnecessary delay. If the mare is unable to assist in regaining her feet, a sling is required. Usually little else is necessary and after a few days in the sling the subject can get about unassisted. In the meanwhile the well-being of the affected animal is to be considered just as in any other case where the patient is so confined. The foal in such instances constitutes a source of some trouble, but the average mare offers no serious resistance to the confinement occasioned by the sling.

Good hygienic care, a suitable diet and full physiological doses of strychnin are indicated. Cadiot and Almy recommend vaginal douches of cold water and counterirritation of the region of the inner thigh in these cases.

Paralysis of the Sciatic Nerve.

Anatomy.—The great sciatic nerve leaves the pelvis in company with the gluteal nerves, through the great sciatic foramen (notch), passing downward along the posterior face of the femur. Near the stifle it passes between the two heads of the gastrocnemius muscle and continues as the tibial. Branches supply the following muscles—obturator, semimembranosus (adductor magnus), biceps femoris (triceps abductor femoris), semitendinosus (biceps rotator tibialis), lateral extensor (peroneus) and the tibial nerve, its continuation, innervates the digital flexors.

Etiology and Occurrence.—Paralysis of the great sciatic nerve may be caused by central disorders, injury in falling, fractures and new growths. Because of its protected position, this nerve does not often suffer injury, and paralysis of the sciatic nerve is recorded in a few instances owing to its rarity.

Symptomatology.—When consideration is given the number of muscles that are supplied by the sciatic nerve and the function of these muscular structures, it is obvious that the leg cannot be used in sciatic paralysis. However, the limb is capable of sustaining weight when it is fixed in position, but this is done without exertion of muscular fibers which are supplied by the great sciatic nerve. Trotting is impossible and flexion of the affected member is also likewise precluded. The foot is dragged when the subject is caused to advance.

Under the heading "sciatica," Scott[44] has described a case of acute sciatic affection wherein a pacing horse manifested evidence of great pain of a nervous character. There were muscular twitchings and the leg was held off the floor and moved about convulsively. Breathing was very much accelerated, pulse 85 per minute, the temperature was 103° and manipulation of the hips augmented the pain.

This was not a paralytic condition and recovery resulted, yet undoubtedly this was a case which, if not properly cared for, might have terminated unfavorably.

Treatment.—Prognosis is decidedly unfavorable in paralysis of the great sciatic nerve. If treatment is attempted, it is to be conducted along the same general lines as in femoral paralysis. Particular attention should be given to conditions which will make for the patient's comfort, and as soon as it is evident that the affection is not progressing favorably, the subject should be humanely destroyed.

Iliac Thrombosis.

This condition is undoubtedly of more frequent occurrence than we are wont to grant when one considers the comparatively small number of cases that are actually recognized in practice. It does not follow, however, that iliac thrombosis rarely exists. Probably in the majority of instances there is insufficient obstruction of the lumina of vessels to provoke noticeable inconvenience. Or, if circulation is hampered to the extent that function is impaired and manifestations are observed by the driver, the subject may be permitted to rest a few days and partial resolution occurs, so that further trouble is not noticeable.

As judged by lesions of the aorta and iliac arteries in dissecting subjects, the conclusion that arteritis and resultant disorders are of rather frequent occurrence, is logical.

Etiology.—Inflammation of the vessel walls and resultant prolifieration of tissue together with the accumulation of clotted blood becoming organized, serve to obstruct the lumen of the affected artery. The cause of arteritis is unknown in many instances, but parasitic invasion and contiguous involvement of vessels in some inflammatory injuries are etiological factors.

Symptomatology.—A characteristic type of lameness signalizes iliac thrombosis and the following brief abstract from a contribution on this subject by Drs. Merillat[45], clearly portrays the chief symptoms:

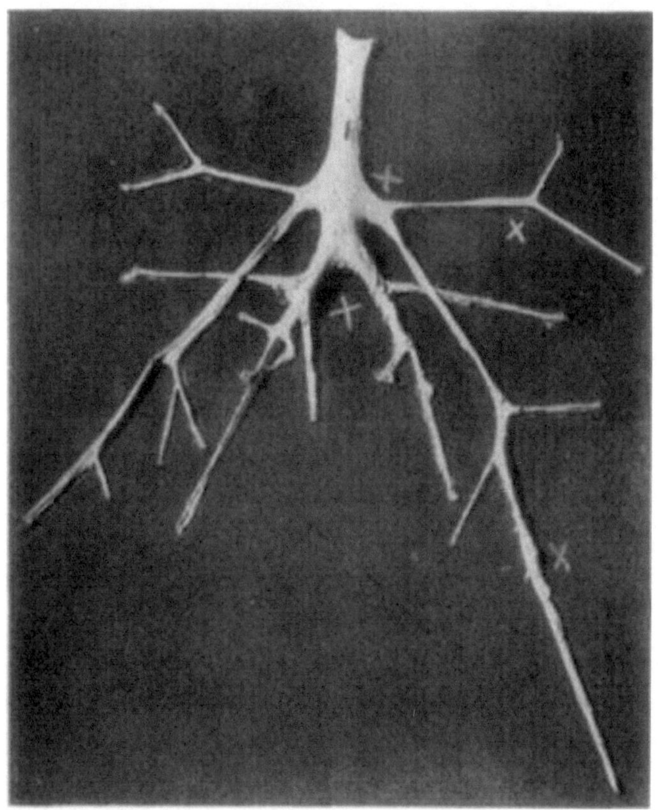

Fig. 48—Exposure of aorta and its branches, showing location of thrombi in numerous places. In this case (same as Fig. 49) Dr. L.A. and Dr. Edward Merillat found the cause of the condition to be due to sclerastomiasis.

The seizures are accompanied with profuse sudation, tremors, dilated nostrils, accelerated respirations and other symptoms of pain and distress, all of which, together with the lameness, disappear as rapidly as they had developed, leaving the animal in an apparently perfect state of health, ready to fall with another attack of precisely the same kind, as soon as enough exercise is forced upon it. The rectal explorations may reveal a pulseless state of one or more of the iliac arteries and a hardness and enlargement of the aortic quadrifurcation, but sometimes this palpation fails to disclose any *percepti-ble* diminution of the blood current of these vessels. The obturation being incomplete, it may be impossible by palpation to decide that thrombosis really exists. In this event and, in fact, in all eases, the clinical symptoms are sufficiently characteristic to make a diagnosis without reservation. It cannot be mistaken for any other disease, once properly investigated. Any given seizure may easily be mistaken for azoturia, at first, but a better examination soon excludes that disease.

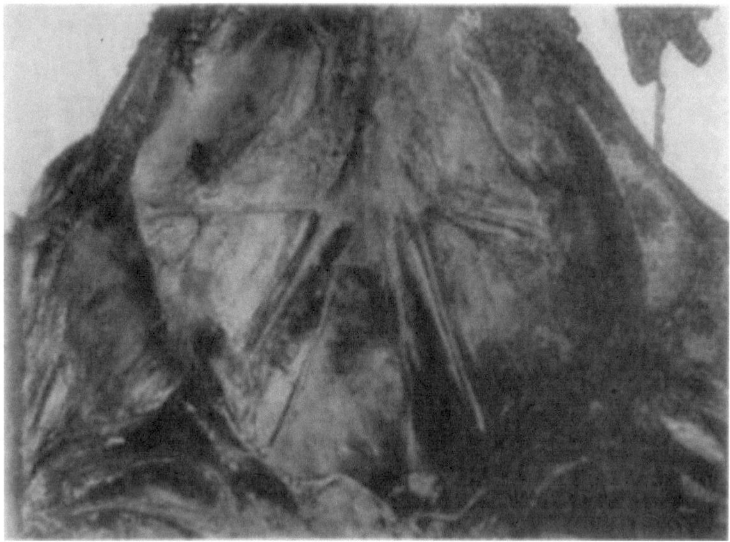

Fig. 49—Illustrative of thrombosis of the aorta, iliacs and branches. Photo by Dr. L.A. Merillat.

Prognosis and Treatment.—In the majority of instances, when there is occasioned serious inconvenience, the outcome is not likely to be favorable, according to Möller. Detachment of a portion of the thrombus, according to Hoare, may result in the lodgment of an embolus in the brain or kidneys. The latter authority also states that muscular atrophy may occur owing to lack of blood supply in some of these cases. Möller states that moderate exercise or work stimulates the establishment of collateral circulation. Massage per rectum is condemned as dangerous by Cadiot.

Fracture of the Patella.

Etiology and Occurrence.—Patellar fractures are rarely met with in the horse but may be caused by falls and heavy contusions. Violent muscular contraction, it is said, may also bring about the same condition.

Symptomatology.—Fracture may be transverse or vertical, and depending on the manner in which the bone is broken, prognosis is either at once rendered favorable or unfavorable. The patella performs a function which is in a way similar to that of the sesamoids and when fractured, complete recovery is improbable in the average instance. When complete, transverse fractures permit of separation of the parts of bone. Tension on the straight ligaments below and contraction of the quadriceps above usually cause insuperable difficulty in the handling of this type of fracture in the horse.

Compound fractures as well as multiple or comminuted fractures occasionally occur and these constitute injuries which are generally considered fatal, although Andrien, according to Cadiot and Almy, succeeded in obtaining complete recovery in a case of compound fracture of the patella and the horse was in service and almost free from lameness two months after treatment was begun.

No difficulty is encountered in recognizing the fracture of the patella because of the exposed position of the bone. Crepitation, and in some cases fissures, may be easily detected.

Treatment.—In simple fracture, when treatment is thought advisable, the subject is put in a sling and kept as nearly comfortable as possible. If little inflammation exists, the application of a vesicant two or three weeks after the injury has been inflicted will be helpful and serve to hasten repair.

Bandages or mechanical appliances are of no practical use in the handling of these cases.

Luxation of the Patella.

Etiology and Occurrence. — This, the most common luxation met with in the equine subject, has been described by writers as existing in many forms. Patellar disarticulation may be more practically considered as *momentary* and *fixed*, regardless of the position taken by the patella. Described under the title of false luxation are recorded cases wherein the quadriceps (crural) muscles become contracted in such manner that a condition simulating true disarticulation of the patella obtains. Also, some practictioners report cases of patellar luxation and refer to pseudo-luxations, without clearly defining the conditions which constitute pseudo-luxation. This has contributed to the extant cause of misconception as to actual differences between luxation and conditions simulating dislocation.

Luxation of the patella is a condition wherein the articular portions of the femur and patella assume abnormal relations whether such displacement of the patella be momentary and capable of spontaneous reduction, or fixed and requiring corrective manipulation. Spasmodic contraction of the crural muscles which sometimes retains the patella in such position that the leg is rigidly extended, does not in itself constitute luxation of the patella; and unless this bone becomes lodged on the upper portion of a femoral condyle or laterally displaced out of its femoral groove, luxation cannot be said to exist in the horse. These are sub-luxations.

Occasionally one may observe in suckling colts outward luxation of the patella wherein there is history of navel infection and no marked evidence of rachitis is present. Some of these cases recover. In a unilateral involvement of this kind in a three-month-old mule colt, the author observed a case wherein an unfavorable prognosis was given and destruction of the subject advised, because of the extreme dislocation of the patella. This colt, however, was not destroyed and in three weeks had apparently recovered. No treatment was given in this instance; the colt was allowed the run of a small pasture with its dam and in time it matured, becoming a sound and serviceable animal.

Classification. — Two forms of true patellar luxation in the horse may be considered; one which is due to the patella becoming fixed upon the internal trochlear rim of the femur and the other when the patella slips over the outer rim of the trochlea.

The first form is known as *upward* luxation and is made possible by rupture of the mesial (internal) femeropatellar ligament. According to Cadiot and Almy, it is only by the rupture of this ligament—the femeropatellar—that upward luxation may occur. This type of luxation is rarely observed and is usually due to violent strain and abnormal extension of the stifle joint.

The second class, *outward* luxation, occurs in colts and is, in many instances, congenital. This form of luxation is also the one usually seen following debilitating diseases such as influenza and pneumonia.

Upward luxation of the patella is characterized by the stiff-extended position of the leg. When the patella is situated upon the inner trochlear rim, the tibia must be extended because of the traction exerted by the straight ligaments. Since the stifle and hock joints extend and flex in unison, there is presented also an extension of the tarsus. Extension of the stifle joint would increase the distance between the femoral origin of the gastrocnemius and its insertion to the summit of fibular tarsal bone (calcis) were it not for the gastrocnemius and superficial flexor (perforatus). Extension of the hock in upward luxation of the patella, permits of flexion of the phalanges. In upward luxation, then, the leg is extended as if too long, but the phalanges may be in a state of moderate flexion. If the foot rests on the ground when the extremity is not flexed, it is almost impossible for the subject to step backward. Because of immobilization of the stifle and hock joints in upward luxation, the subject can walk only by hopping on the sound leg and then the extremity is flexed, allowing the anterior portion of the fetlock to drag on the ground.

In some cases practitioners are called to attend young animals that are reported to be "stifled" (often in young mules that have made a rapid growth) and upon arrival the only noticeable symptom of preëxisting luxation is the soiled condition of the anterior fetlock region—evidence of its having been dragged. Such cases may be styled momentary luxation, whether they are due to a weakened condition of the patellar ligaments or spasmodic contraction of the crural muscles.

In upward luxation, reduction is effected by attempting further extension of the stifle joint and at the same time the patella is pulled outward, off the internal rim of the trochlea. This is attempted by securing the subject in a standing position; the sound side is kept against a wall if possible and a rope is tied to the extremity of the affected leg. Traction is exerted upon the rope and at the same time force is directed against the stifle joint to produce further extension if possible, so that the straight patellar ligaments may relax sufficiently to allow the patella to be dislodged from its position upon the inner trochlear lip. Failing in this manner of procedure, the affected animal is to be cast and anesthetized with chloroform. The relaxation which attends surgical anesthesia will permit of reduction of the dislocated bone and manipulations such as have just been outlined may be employed.

Following reduction in the average case it is essential that the subject be given vigorous exercise for a few minutes. Reduction having been affected, the application of a vesicant over the whole patellar region is customary.

In cases of habitual luxation, unless the ligaments are so lax that the patella may be displaced laterally over the inner as well as the outer trochler rims, division of the inner straight patellar ligament will correct the condition. This desmotomy has been advocated by Bassi, and good results in appropriate cases have been reported by Cadiot, Merillat and Schumacher. This operation has been found a corrective in cases of outward luxation as well as those of upward dislocation of the patella when resorted to before the trochleae are worn from frequent luxation.

Outward luxation of the patella is occasioned by a lax condition of the internal femeropatellar ligament or a rupture of the same so that the patella slips over the outer femoral trochlear rim and permits of an abnormal flexion of the stifle joint. The outer trochlear rim being the smaller of the two, inward luxation does not occur in the horse. With the patella disarticulated in this manner, the action of the quapriceps femoral group of muscles has no effect on the stifle joint and, therefore, flexion of this articulation occurs as soon as the subject attempts to sustain weight and the leg collapses unless weight is at once taken up by the other member if sound.

As a rule, the reduction of this form of luxation is not difficult. The patella may be pushed inward and into position without manipulation of the leg. Retention of the patella in position is a difficult problem. Bandaging is considered impractical and is not ordinarily done in this country. Benard, according to Cadiot and Almy, recommends bandaging with a heavy piece of cloth in which an opening is made through which the patella is allowed to protrude, and by turning such a bandage snugly about the stifle several times, the patella is held in position. This bandage should be kept in place for about ten days.

In young and rachitic animals outdoor exercise and a good nutritive ration for the subject are indicated. Hypophosphites in assimilable form may be beneficial, and vesication of the patellar region contributes to recovery.

Where extreme luxation is present in both stifles, the prognosis is unfavorable. In such cases, degenerative changes may exist and in some instances the ligaments are so diseased and elongated that regeneration is impossible. Williams[46] reports a case where bilateral "floating" (outward) luxation was present and extensive degeneration changes affected the articulation.

In subjects suffering frequent dislocation of the patella (habitual luxation) it is possible in some cases, to prevent its occurrence or at least to minimize the distress occasioned by momentary luxation, by keeping the animals in wide stalls so that "backing" is unnecessary. In some nervous subjects that seem to be suffering from cramp of the crural muscles, the difficulty and pain of their being backed out of narrow stalls, accentuates the nervousness. Sudation and restlessness are manifested and the subject presents a clinical picture of distress and fear of a painful ordeal. In some cases of this kind, complete recovery takes place by the time animals are five or six years of age. One should avoid keeping such subjects in narrow stalls. Preferably patellar desmotomy should be performed that relief may be obtained at once.

Luxations attending some cases of influenza recover promptly when subjects are kept comfortably confined in roomy box-stalls. The administration of stimulative medicaments such as nux vomica and the application of an active blistering agent to the patella serve to hasten recovery. Dislocations in such cases are often bilateral and they are usually momentary. Reduction occurs spontaneously, as a rule, and the subjects are not occasioned much distress if they are kept quiet for a few days.

Chronic Gonitis.

Etiology and Occurrence.—Chronic inflammation of the stifle joint is met with following acute synovitis due to strains and concussion. It is an ailment which affects heavy horses and particularly animals that are kept at work on paved streets, but this does not explain its existence in animals that are not subjected to work likely to cause concussion. Berns[47] considers rheumatism a probable cause of gonitis and, as he states, the dropsical form of affection of this joint is not ordinarily attended with manifestations of inconvenience to the subject. Gonitis is often bilateral and its onset is insidious in many instances.

Symptomatology.—In unilateral gonitis weight is not borne by the affected member. There is noticeable distension of the joint capsule—a characteristic pendant pouching protrusion. When both stifles are affected the subject frequently shifts the weight from one limb to the other. Lameness comes on gradually and during the incipient stages may be intermittent but it progressively increases so that in time affected animals become useless. In bilateral affections animals drag the toes because of the pain incident to flexing the stifles. This is particularly evident when the subject is made to trot. As the disease progresses, atrophy of the quadriceps femoris muscles becomes pronounced and as destructive changes involving the articular cartilages take place. The subject becomes more lame and eventually is rendered incapable of service.

Upon manipulation of the patellar region, one is impressed with the fact that hyperesthesia does not exist in proportion to the pain manifested during locomotion. In some cases a gelatinous swelling is present and may be detected by palpating between the straight ligaments of the patella. Williams, Hughes, Merillat, Hadley and others have directed attention to the existence of floating masses (*corpora oryzoidea*) in the synovial capsule of this joint in gonitis, and as with all cases of arthritis, irreparable damage is often done the articular cartilages during the course of the ailment.

Fig. 50—Chronic gonitis. The knuckling which results from long continued inactivity of the crural muscles in chronic cases is marked in this instance. Photo by Dr. L.A. Merillat.

Treatment.—No effective method is as yet known which will control this condition during its incipiency. The disease progresses, and more or less damage is done the affected parts in the course of months or even years in some cases before subjects are rendered hopelessly crippled. When recognized early (before chronic gonitis exists) aspiration of the synovia and the injection of diluted tincture of iodin might prove beneficial in cases of synovial distension. Chronic gonitis is considered an incurable affection and as soon as subjects manifest evidence of distress from this condition they should by all means be taken from work. Firing and vesication have not been productive of beneficial results.

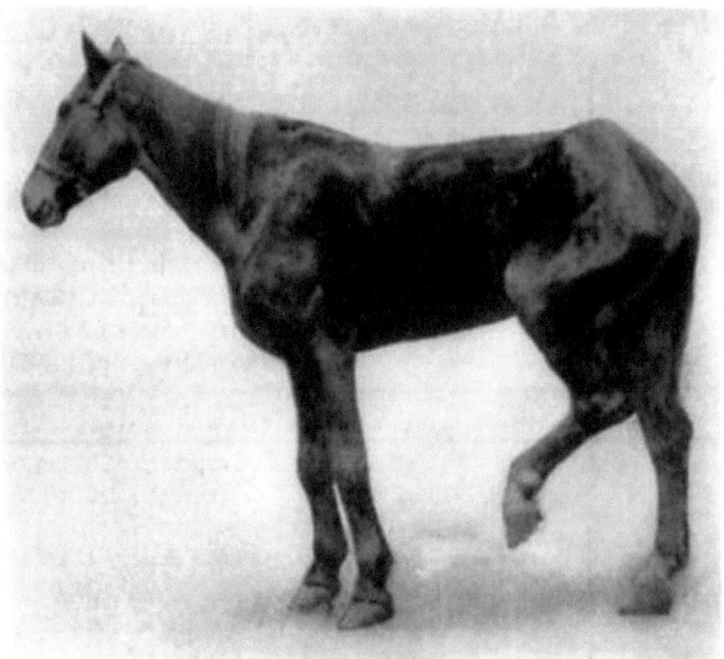

Fig. 51—Gonitis. Showing position assumed in such cases because of pain occasioned. Photo by Dr. C.A. McKillip.

Open Stifle Joint.

Anatomy of the Joint Capsule.—This joint capsule is thin and very capacious. On the patella it is attached around the margin of the articular surface, but on the femur the line of attachment is at a varying distance from the articular surface. On the medial side it is an inch or more from the articular cartilage; on the lateral side and above, about half an inch. It pouches upward under the quadriceps femoris for a distance of two or three inches, a pad of fat separating the capsule from the muscle. Below the patella it is separated from the patellar ligaments by a thick pad of fat, but inferiorly it is in contact with the femerotibial capsules. The joint cavity is the most extensive in the body. It usually communicates with the medial sac of the femerotibial joint cavity by a slit-like opening situated at the lowest part of the medial ridge of the trochlea. A similar, usually smaller, communication with the lateral sac of the femerotibial capsule is often found at the lowest part of the lateral ridge. (Sisson's Anatomy.)

Thus it is seen that because of its frequent communication with the other parts of this large synovial membrane, a wound which opens the external portion of the femerotibial capsule may be the cause of contamination and resultant infectious arthritis of the whole stifle joint. Because of the distance between the most dependent part of the femerotibial articulation and the summit of the patella, one may misjudge the exact location of the lowermost part of this portion of the capsular ligament of the stifle joint and thereby fail at once to appreciate the seriousness of calk wounds in this region.

Etiology and Occurrence.—Wounds to the patellar region are of rather frequent occurrence, and because of the comparatively unprotected position of these structures, the capsular ligaments of the stifle joint may be perforated as a result of violence in some form. Calk wounds which penetrate the tissues in the immediate region of the lower portion of the external part of the femerotibial capsule sometimes result in open joint because of tissue necrosis resulting from the introduction of infection. Contused wounds sometimes destroy the skin and fascia over large areas on the lateral patellar region and because of subsequent sloughing of tissue due to infection as well as to the manner in which such wounds are inflicted, septic arthritis subsequently occurs. Penetrant wounds, such as may be caused by a fork tine may not result in infection; if infectious material is introduced an infectious arthritis does not necessarily follow, though such cases should be considered as serious from the outset.

Symptomatology.—The pathognomonic symptom of open stifle joint is the profuse escape of synovia, indicating perforation of the synovial capsule; by means of a probe the wound may be explored in a way that will clearly reveal the nature of the injury.

After a few days have elapsed in cases where considerable infection has taken place, there is manifestation of pain as in all cases of infective arthritis. Hughes[48] gives an excellent description of the clinical aspect of arthritis which applies here:

Acute arthritis begins like an ordinary attack of synovitis. In joints other than the pedal and pastern, there is sudden and extensive swelling, which at first is intra-articular, succeeded by extra-articular tumefaction, and accompanied by violent lameness. The pain soon becomes intense and agonizing. There is severe constitutional disturbance, the temperature ranging from 104 to 106 degrees and the pulse from 60 to 72. Painful convulsions of the limb occur, shown by involuntary spasmodic elevations due to reflex irritation of the muscles. There is loss of appetite, rapid emaciation, the flank is tucked up and the back arched. In from three to six days, the tumefaction around the joint tends to soften at a particular place, and bursts, and a discharge that is sometimes of a sanious character, mixed with synovia, escapes. Great exhaustion at times supervenes, and if the joint is an important one, the horse lies or falls and is unable to rise.

Treatment.—In small puncture wounds the immediate application of a vesicating ointment has given good results, but when infection has taken place to such extent that the animal manifests evidence of intense pain, and lameness is marked and local swelling and hyperesthesia are great, vesication is contraindicated. In such instances the exterior of the wound and its margins should be prepared as in similar affections of other joints. A quantity of synovia is then aspirated by means of a small trocar and care should be taken to observe all due aseptic precautions. Subsequently the injection of from four to six ounces of a mixture of tincture of iodin, one part to ten parts of glycerin, and gentle massage of the joint immediately after the injection has been made, serves to check the infective process in some cases.

The subject should be cared for as has been previously suggested in arthritis proper provisions for comfort being made. Good nursing is always essential to a successful issue. However, the author cannot view cases of open stifle joint with the same optimism concerning their course and outcome that is expressed by a number of writers on this subject. It is a grave condition wherein the prognosis should be given advisedly.

Fracture of the Tibia.

Etiology and Occurrence.—Because of its exposed position to kicks, and its lack of protection by heavy musculature (especially on its inner surface), there is afforded ample opportunity for frequent injury to the tibia. Fractures are complete and varying as to nature, or incomplete. The heavy tibial fascia affords sufficient protection so that fissures without entire solution of continuity of the bone may occur from violence to which this part is often subjected. Möller classes tibial fracture as ranking second in frequency—pelvic fracture being more often met with in horses. This does not apply in our country as phalangeal and metacarpal and even metatarsal fractures are observed in more instances than are such injuries to the tibia. The tibia is occasionally broken at its middle and lower thirds, but malleolar fractures are not common.

Symptomatology.—When fracture is complete and all support is removed, the leg dangles, and the nature of the injury is so obvious that there is no mistaking its identity. However, in case of incomplete fracture one needs to base all conclusions upon the history of the case, evidence of injury, or other knowledge of the character of violence to which this bone has been exposed. For without the presence of crepitation (even by excluding other possible causes for the pronounced lameness which characterizes some of these cases) we can only resort to the knowledge which experience has taught that fracture may be deemed probable in many injuries to the tibial region. Consequently, we are to look upon all injuries that affect the tibia as being fractures of some sort when there is either local evidence of the infliction of violence or whenever marked lameness attends such injuries, unless there is positive indication that no fractures exist.

A careful examination of parts of the tibia, i.e., noting the amount and painfulness of swellings, exploration with the probe, and observations of the course taken in any given case, will determine the exact nature of injuries. Such examination needs to extend over a period of a week or in some instances two or three weeks may pass before the true state of affairs is apparent. In the meanwhile, cases are to be handled as though tibial fracture certainly existed.

Prognosis.—Prediction of the outcome in tibial fracture is somewhat presumptuous, but in the majority of cases in mature subjects fatality results. Cadiot[49], however, views this condition with more optimism than have American practitioners. While he considers the condition grave, in citing case reports of successful treatment by d'Arboval, Duchemin, Leblanc, and others, his conclusion is that many practitioners erroneously consider fractures of the tibia as incurable.

The method of handling these cases by Leblanc is as follows: The subject is placed in a sling; a pit is excavated below the affected member so that a heavy weight may be attached to the extremity; splints are applied to each side of the leg, which is padded with oakum, and this is kept in position by means of bandages covered with pitch. The outer splint extends from the hoof to the stifle and the inner one from the hoof to the upper third of the leg. This method in the hands of Leblanc has been successful in several instances, according to Cadiot.

In a foal the author has in one instance succeeded in obtaining complete recovery in a simple fracture of the lower third of the tibia where the only support given the broken bone was a four-inch plaster-of-paris bandage which was adjusted above the hock. Below the tarsus a cotton and gauze bandage was applied to prevent swelling of the extremity. In this instance (an emergency case in which materials that are not to be recommended were necessarily employed) recovery took place within thirty days.

As has been mentioned in the consideration of radial fractures, heavy leather is better suited for immobilization of these parts than a cast or other rigid splint materials. Mature animals may be expected to resist the immobilization of the hind legs because of the normal manner of flexion of the tarsal and stifle joints in unison. Therefore, the application of rigid splints to the leg and including the hock is productive of disastrous results in some cases.

The application of cotton and bandages to pad the member and the adjusting of heavy leather splints on either side of the leg, and retaining them in position with four-inch gauze bandages will prove more nearly satisfactory than some other methods employed. Prognosis is unfavorable, however, in most cases of compound fracture and recovery is improbable when the upper portion of the tibia is broken.

Rupture and Wounds of the Tendo Achillis.

Etiology and Occurrence.—Cases are recorded by Uhlrich in which rupture has followed degenerative changes affecting the tendo Achillis. Not infrequently, the result of a trauma, division of the tendo Achillis occurs. Möller states that rupture of this tendon may be due to jumping, in riding horses and in draught horses, in their efforts to avoid slipping. In runaways, it sometimes occurs where sharp-edged implements are bounced against the legs in such fashion that division of the tendon results.

Symptomatology.—With division of the tendo Achillis or of the musculature of the gastroenemii and the superficial flexor (perforatus), there remains nothing to inhibit tarsal flexion except the deep flexor tendon (perforans) and this does not support the leg. When attempt is made to sustain weight with the affected member, abnormal flexion of the tarsus takes place and the hock sinks almost to the ground. The symptoms are so characteristic that recognition is always easy even in case no wound of the skin exists.

Prognosis.—Spontaneous recoveries occur and such cases are reported by Bouley who is quoted by Cadiot as having observed division of the tendo Achillis due to a sword wound wherein at the end of four months recovery was complete. Division of this tendon in brood mares has been practiced by the early settlers of parts of the United States for the purpose of preventing their straying too far from home. In such instances one leg only was so mutilated and in most instances, it is reported that spontaneous recovery took place.

In unilateral involvement without complications, the prognosis is not unfavorable if provisions for giving necessary attention are available.

Treatment.—The subject is to be confined in a sling and the member bandaged and supported by means of leather splints. Immobilization as for fracture is not necessary but, nevertheless, movement is to be restricted as much as possible. In case of open wounds, the exposed tissues are cared for along general surgical lines. Where the divided parts of the tendon are maintained in fairly close and constant relation, granulation of tissue, sufficient to sustain weight takes place in from six weeks to three months.

Spring-Halt. (String-Halt.)

Occurrence.—This condition is a myoclonic affection of the hind leg which is discussed in works on theory and practice under the head of neuroses, but the cause or causes have not been established. Theories that heredity is responsible have their supporters and advocates of hypotheses attributing it to disease of the sciatic nerve, patellar subluxation, fascial contraction of various muscles, "dry spavin" (tarsal arthritis), iliac exostoses, disease of the foot and contraction of the hoof, are on record in veterinary literature. This ailment affects old horses more frequently than it does young and is seen in all breeds of animals including mules.

Fig. 52—Spring-halt.

Symptomatology.—This disease develops slowly, and progressively increases in severity as a rule, but does not ordinarily constitute cause for rendering an animal unserviceable. While the affection is sometimes bilateral (occasionally affections of the forelegs are reported) and the extreme flexion of the legs in the spasmodic manner which characterizes spring-halt, cause great waste of energy during locomotion, yet such cases are rare. Usually the ailment is markedly evinced when subjects are first taken from the stable, but as they are exercised the manifestation diminishes, and in many instances it completely subsides. The condition is generally more noticeable when the subject is made to step backward. In some animals there is marked abduction at the time flexion occurs and in singular instances the spasmodic contraction is so violent that the subject falls to the ground as a result of the peculiar flexion of the leg.

In severe cases of "scratches" or chemical irritation of the extremity, the legs are abnormally flexed in a manner which simulates spring-halt, but because of the evident injury of the parts this is not likely to confuse. Since all facts concerning etiological agencies are surrounded with so much obscurity, classification does not lend any particular assistance in the consideration of this ailment.

Prognosis.—One cannot intelligently give a prognosis in these cases if forecast is expected to state the exact course following treatment. However, in a general way, cases of recent affection are thought more favorable than are those of long standing or in old animals where myositis and other muscular and fascial affections exist owing to years of hard service.

Treatment.—No known line of medicinal treatment is of service, nor is any particular surgical operation to be considered dependable for obtaining relief. Operations of almost every conceivable nature have been tried with the hope of securing recovery in spring-halt but under no condition can the practitioner as yet be reasonably certain of effecting permanent relief in any case. Treatment is, therefore, entirely empirical.

Neurectomies have been performed and recoveries following were attributed thereto; fascial divisions in the crural region have been done with good results and this manner of treatment has its favorers. Advocates of tenotomies, likewise, are to be found. Consequently, one may summarize thus: Spring-halt is a disease of unknown origin—the exact cause has not been determined; therefore, all treatment is, in a way, experimental. The recommendation of any given procedure in handling cases must then be a matter of opinion based either upon practical experience or knowledge of the experiences of others. Divisions of the lateral digital extensor (peroneus) below the tarsus near its point of insertion to the extensor of the digit is recommended here because it is followed by a percentage of recoveries that is as large as in any other method of treatment and the operation is not difficult to perform nor is its performance fraught with any dangerous complications. In selected subjects about fifty per cent of cases recover in from two to six weeks following this operation.

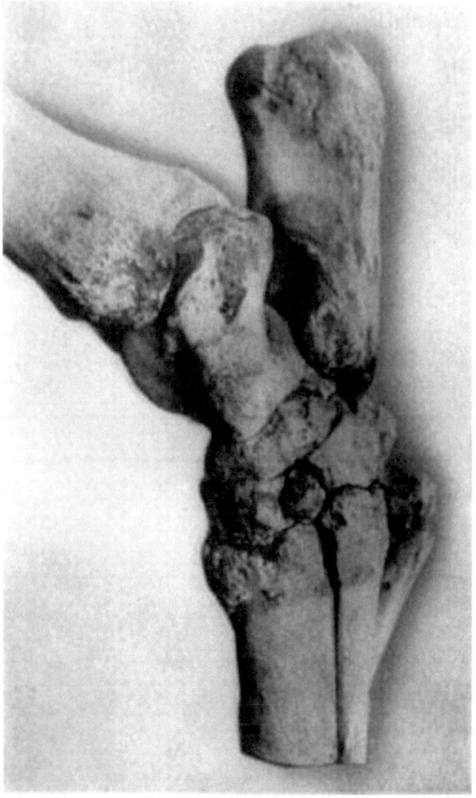

Fig. 53 — Lateral (external) view of tarsus showing effects of generalized tarsitis.

Open Tarsal Joint.

Like the tibia the hock is exposed to frequent injuries and in some cases wounds perforate the joint capsule. When due to calk wounds where horses are kicked, the injury is often on the side of the tarsus (medial or lateral) and such wounds not infrequently result in infectious arthritis. Horses sometimes jump over wire fences and wounds are inflicted which constitute extensive laceration of the joint capsule. In firing for bone spavin, where a deep puncture is made very near the tibial tarsal (tibioastragular) joint if infection gains entrance, serious and generalized infection of the open joint cavity supervenes in some cases.

Symptomatology.—There is no marked difference in the constitutional disturbances which are occasioned in this condition and those encountered in other cases of septic arthritis (previously considered herein) except that there is a difference in the degree of resultant derangement and local tissue changes. Chiefly, because of the difficulty encountered in keeping the hock joint in an aseptic condition or securely bandaged, open tarsal joint constitutes a more serious condition than a similar affection of the fetlock. Otherwise, a very similar condition obtains and the same diagnostic principles serve here that have been described on page 110 in considering open fetlock joint.

Treatment.—The same plan that is described in detail for treatment of similar conditions affecting the fetlock joint is indicated in this affection. Exceeding care must be exercised in bandaging the hock, however, lest the animal be so irritated that in the extreme flexion of the tarsus which is often caused by bandaging, the wound dressings may be completely deranged. A wide gauze bandage material is most satisfactory; cotton of long fiber is separated in thin layers and wound about the hock, extending from the site of injury to a point about six inches proximal to the summit of the os calcis. By using an abundance of cotton in this way, it will not be found necessary to apply the bandages very snugly; with a four-inch gauze bandage material, which is supported above the cap of the hock and brought across the anterior face of the tarsus in a diagonal manner, a comfortable and very serviceable protective dressing is provided for. Animals so treated will not ordinarily resist because of pressure from the bandages. Pressure is unavoidable in the use of adhesive dressings or where careful attention is not given the manner of applying cotton to the parts. Such methods are sure to result disastrously. But if subjects are kept quiet after the parts have been properly bandaged, no difficulty is encountered in maintaining asepsis in an uninfected wound. Recovery takes place in favorable cases in from three weeks to three months, depending on the nature and extent of injuries inflicted.

Fracture of the Fibular Tarsal Bone (Calcaneum.)

Etiology and Occurrence.—This condition though rarely met with in the horse, is the result of violent strain upon the os calcis by the gastrocnemius and superficial flexor tendons in efforts put forth by animals in attempts to regain a footing when the hind feet slip forward under the body, or in jumping and in falls or direct contusion by heavy bodies. Hoare[50] reports a case of a mare that had produced fracture in jumping.

Fracture of the other tarsal bones are very seldom observed but may be occasioned by contusions wherein multiple or comminuted fractures are produced, such as are to be seen in small animals. Fracture of the tibial tarsal bone (astragalus) is to be observed as a complication in luxations of the tarsal joint and, according to Cadiot, the other tarsal bones may likewise suffer fracture in luxations of the hock.

Symptomatology.—Great pain attends this accident according to the observations given in recorded cases. In the case cited by Hoare the animal evinced great pain and uneasiness; the hock was unduly flexed; the calcaneum was displaced forward; and marked crepitation was present. A portion of the body of the calcaneum was protruding through the perforated skin. The animal was destroyed and the bone was found broken in three pieces.

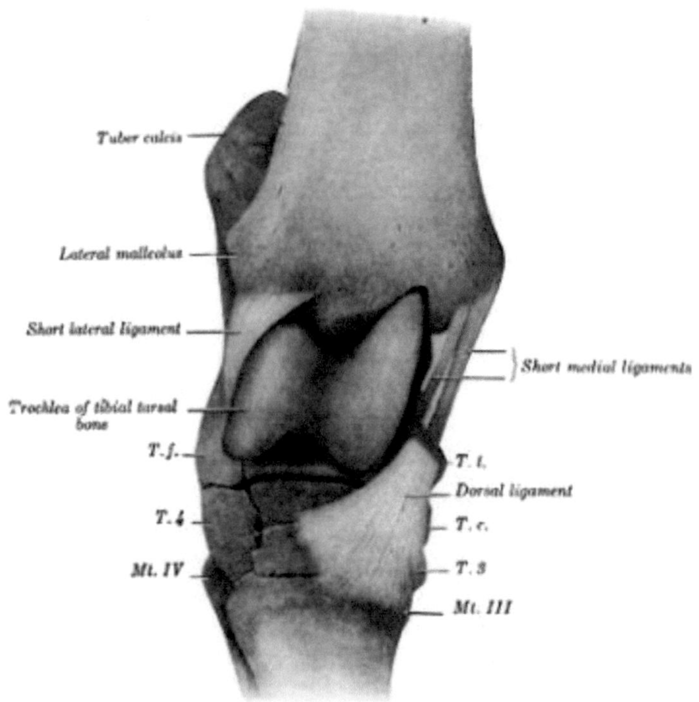

Fig. 54—Right hock joint. Viewed from the front and slightly laterally after removal of joint capsule and long collateral ligaments. T. t., Tibial tarsal bone (distal tuberosity). T. c., central tarsal bone. T. 3. Ridge of third tarsal bone. T. f. Fibular tarsal bone (distal end). T. 4. Fourth tarsal bone. Mt. III, Mt. IV. Metatarsal bones. Arrow points to vascular canal. (From Sisson's "Anatomy of the Domestic Animals.")

Since the support for the tendo Achillis is removed in such fracture and no leverage on the metatarsus obtains, it naturally follows that any attempt to sustain weight must result in extreme flexion of the hock and descent of this part in a manner similar to cases of rupture or division of the Achilles' tendon. The two conditions should not be confused, however, as the parts may be definitely outlined by palpation and the slack condition of the tendon and displaced summit of the calcaneum, which characterize fracture of the fibular tarsal bone, are easily recognized.

Treatment.—Prognosis is unfavorable in the majority of cases, but should attempts at treatment be undertaken in young and quiet mares which might prove valuable for breeding purposes in case of imperfect recovery, they should be put in slings and the member is to be immobilized as in tibial fracture. Authorities are agreed that prognosis is entirely unfavorable in mature animals, when the case is viewed from an economic standpoint.

Tarsal Sprains.

Etiology and Occurrence. — The hock joint is often subjected to great strain because of the structural nature of this part and its relation to the hip as well as the manner in which the tarsus functionates during locomotion. That ligamentous injuries owing to sprain frequently occur and attendant periarticular inflammations with subsequent hypertrophic changes follow, is a logical inference. Fibrillary fracture of the collateral ligaments may take place in falls or when animals make violent efforts to maintain their footing on slippery streets. In expressing opinions concerning the frequency with which the hock is found to be the seat of trouble in lameness of the pelvic members, different writers place the percentage of hock lameness at from seventy-five to ninety per cent. And when one considers the possibility that a goodly proportion of cases of tarsal exostis are the outcome of sprains, the occurrence of tarsal sprains may be more generally admitted.

Symptomatology. — A mixed type of lameness is present and the nature of the impediment varies, depending upon the location of the injury. Sprains of the mesial tarsal ligaments cause lameness somewhat similar to that of spavin. However, in establishing a diagnosis, local evidence in these cases is of greater significance than the manner of locomotion. During the acute stage of inflammation there is to be detected local hyperthermia, some hyperesthesia and a little swelling. Later, when resolution is not prompt, considerable swelling (or perhaps correctly speaking, an indurated enlargement) variable in size is developed. In some cases the entire tarsal region becomes greatly enlarged and this swelling is very slowly absorbed in part or completely. Such sub-acute cases are observed during the winter season and particularly where subjects are kept in tie stalls without exercise for weeks at a time.

Treatment. — Attention should be directed toward relief for the animal in all acute inflammations. Local applications of heat are helpful and, of course, rest is essential. Towels that are wrung out of hot water and held in position by means of a few turns of a loose bandage and this covered with an impervious rubber sheet, will serve as a practical means of application of hydrotherapy. Following this when conditions improve, as in the handling of all similar cases, counterirritation is indicated.

When proper care is given at the onset and where injury does not involve too much ligamentous tissue, recovery takes place in a few weeks but in some cases which occur during the winter season in farm horses, complete recovery does not result until several months have passed.

Curb.

The hock is said to be curbed when the normal appearance, viewed from the side, is that of bulging posteriorly at any point between the summit of the calcaneum and the upper third of the metatarsus. Among some horsemen a hock is said to be "curby" whenever there exists an enlargement of any kind on the posterior face of the tarsus whether it be due to sprain, exostosis or proliferation of tissue as a result of contusion.

French veterinarians consider under the title of "courbe," an exostosis situated on the mesial side of the distal end of the tibia. Cadiot and Almy state that this condition (courbe) is of rare occurrence. Percivall defines curb as "a prominence upon the back of the hind leg, a little below the hock, of a curvilinear shape, running in a direct line downwards and consisting of infusion into, or thickening of, the sheath of the flexor tendons." Möller's version of true curb is a thickening of the plantar ligament (calcaneocuboid or calcaneometatarsal). Hughes and Merillat consider curb as a synovitis having for its seat the synovial bursa which is situated between the superficial flexor tendon (perforatus) and the plantar ligament.

Occurrence.—Certain predisposing factors seem to favor the occurrence of curb. A malformation of the inferior part of the tarsus so that its antero-posterior diameter is considerably less than normal is a contributing cause. Such hocks are known as "tied-in." Another fault in conformation is the existence of a weak hock that is set low down on a crooked leg, especially when such a member is heavily muscled at the hip. Given such conformation in an excitable horse, and curb is usually produced before the subject is old enough for service. It is certain that in cases where conformation is bad, greater strain is put upon the plantar ligament. This structure serves to bind the tibial tarsal (calcis) bone to the metatarsus; traction exerted upon its summit by the tendo Achillis is great when animals run, jump or rear and also at heavy pulling. In animals having curby hocks, sprain is likely to result and curb supervenes.

Symptomatology.—The characteristic swelling which marks curb may develop quickly and lameness occur suddenly or the enlargement comes on gradually and slowly, causing little lameness.

Lameness is not proportionate to the size of the swelling and in all cases whether subacute or chronic, the condition improves with rest, but lameness is again manifested upon exertion. A horse which "throws a curb" will go lame until the acute inflammatory condition subsides and depending upon treatment received and conformation of the hock, this requires from three days to two or three weeks.

The character of the swelling varies; in some cases it is not large but rather dense and lacking in evidence of heat and hyperesthesia; in other cases there is considerable swelling, which is hot and doughy, somewhat painful to the touch but not necessarily productive of much lameness. In any event, whether the swelling or enlargement is big or little, its location makes it conspicuous when viewed in profile.

In most cases after the acute inflammatory period has passed, lameness is slight, if at all present, and in time no interference with the subject's usefulness is occasioned because of the curb, but the animals often remain blemished—complete resorption of inflammatory products being unusual when much disturbance has existed.

Treatment.—The handling of curb during the acute inflammatory stage is along the same lines as in sprain—local applications of cold and heat. Subjects must be kept quiet until all inflammation has subsided, for there are no cases wherein a little brisk exercise is more likely to cause a recurrence of lameness before recovery is complete than in curb. Vesication is in order in a week or ten days after the affection has set in; in old stubborn cases that have resisted ordinary treatment for a few months, the use of the actual cautery (line firing) is to be recommended.

Fig. 55—Spavin.

Spavin. (Bone Spavin.)

This term is applied to an affection of the tarsus which is usually characterized by the existence of an exostosis on the mesial and inferior portion of the hock. There is also included under this name, articular inflammation wherein no external evidence is shown. Spavin lameness has long been recognized and much has been written upon this subject. Since authorities are agreed that most cases of lameness in the hind leg are due to hock affection, and because the majority of cases of lameness which have the tarsal region as the seat of trouble are instances of spavin lameness, this disease merits all the attention it has received.

Etiology and Occurrence.—Causes may well be classified as predisposing and exciting, for there are many etiologic factors to be reckoned with in spavin, some of which are widely different in nature.

Considered as predisposing causes, hereditary influences play an important rôle and may, owing to faulty conformation, subject an animal to affections of this kind because of disproportionate development of parts (weak and small joints and heavy muscular hips); or as a consequence of inherited traits, a subject may manifest susceptibility to degenerative bone changes which are signalized by the formation of exostoses of different parts on one or more of the legs. Hereditary predispositions make for the presence of spavin in a large percentage of the progeny of sires so affected. This fact has been repeatedly demonstrated in this country as well as elsewhere according to Quitman, Dalrymple and Merillat.[51] A number of states have passed stallion inspection laws stipulating that animals having such exostoses as spavin and ringbone cannot be registered except as "unsound."

Asymmetrical conformation, particularly where the hock is obviously small and weak as compared with other parts of the leg, constitutes a noteworthy predisposing cause.

Peters' theory is plausible that the screw-like joint between the tibia and the tibial tarsal (astragulus) bones causes these structures to functionate in a manner not in harmony with the provisions allowed by the collateral ligaments of the tarsus, permitting movement only in a direction parallel with the long axis of the body.

Because of the quality of their temperaments, nervous animals possessing no particular congenital structural defects of the hock and having no history of spavined progenitors, are subject to spavin when kept at work likely to produce tarsal sprain. Spavin usually develops early in such subjects and examples of this kind may be frequently observed in agricultural sections of the country. Where spavin develops in unshod colts at three and four years of age, shoeing is not an influencing agency when animals are not worked on pavements.

Exciting causes of spavin are sprain and concussion. Various hypotheses are recorded as to how sprains are influenced and among others may be mentioned that of McDonough[52], which is that the foot is robbed of its normal manner of support by the ordinary three-calked shoe. With such a shoe, little support is given the sides of the foot; hence, undue strain is put upon the collateral ligaments of the tarsus. Moreover, the shoe with its calks increases the length of the leg and adds to the leverage on the hock, by virtue of such added length. This makes for greater strain upon the mesial or lateral tarsal ligaments whenever the foot bears upon a sloping ground surface, so that one side (inner or outer) is higher or lower than the other. But according to McDonough's theory (a good one concerning horses that work on pavements), the chief error in shoeing lies in that the foot is deprived of its normal base or support on the sides—the three-calked shoe being an unstable support—and that this manner of shoeing city horses working on pavements is an "inhumane" practice, a "diabolical method."

Whether spavin has its point of origin within the articulation as a rarefying ostitis of the cancellated structure of the lower tarsal bones as suggested by Eberlein; or, as Diekerhoff asserts, that the cunean bursa may be the initial point of affection, is unsettled; but it is reasonable to consider occult spavin as having its origin within the articulation, and that cases readily yielding to cunean tenotomy are primarily due to affection of the cunean bursa.

Symptomatology.—Where a visible exostosis exists, the presence of spavin is easily detected, yet exostoses that extend over large areas may constitute cause for serious trouble and still be difficult of detection. By observing the internal surface of the hock from various suitable angles, such as from between the forelegs or directly behind the subject, one may note the presence of any ordinary exostosis.

The position assumed by the spavined horse is often characteristic. More or less knuckling is usually present (Liautard, McDonald). There is abduction of the stifle in some cases, or the toe may be worn in unshod horses so that it presents a straight line at the surface. This is manifested to a great degree in some animals and in others the foot is not dragged and there is no wearing of the hoof at the toe.

Spavin lameness is so distinctive that one trained and experienced in the examination of horses that are spavined, should correctly diagnose the condition in practically every instance without recourse to other means than noting the peculiar character of the gait of the subject. Lameness develops gradually in the majority of instances, and an important feature in spavin lameness is that it disappears after the subject has gone a little way, to return again as soon as the animal has rested for a variable length of time—from a half hour to several hours. This "warming out" is marked during the incipient stage, but less pronounced in most chronic cases. A complete disappearance of lameness is observed in some instances, while in others only partial subsidence is evident. Because of the fact that pain is occasioned both during weight bearing and while the leg is being flexed and advanced, there is manifested the characteristic mixed lameness and exaggerated hip action which typifies spavin. By throwing the hips upward with the sound member it is possible to advance the affected leg with less flexion, hence less pain is experienced in this manner of locomotion. When made to step aside in the stall, a spavined horse will flex the affected member abruptly and when weight is taken on the diseased leg, symptoms are evinced of pain, and weight is immediately shifted to the sound limb. This is marked during the incipient stages of spavin. Lameness usually precedes the formation of exostosis, though cases are observed wherein an exostosis is present and no lameness is manifested and no history of the previous existence of lameness is available.

The "spavin test" is of value as a diagnostic measure when it is employed with other means of examination, though reaction to this test is seen in some cases in old "crampy" horses that have experienced hard service. The test consists in flexing the affected leg (elevating the foot from the ground twelve to twenty-four inches) and holding the member in this position for a minute, whereupon the animal is made to step away immediately at a trot. During the first few steps taken directly thereafter, the subject shows pronounced lameness and this constitutes a reaction to the spavin test.

Where no exostosis is present it becomes necessary to exclude other causes for lameness but the characteristic spavin lameness is to be relied upon to a greater extent in such cases than are other means of examination. Such cases are known as occult spavin and may be present for months before any external changes in structure are observable. In some instances no extoses form even during the course of years. The spavin test is of aid in establishing a diagnosis here but the marked "warming out" peculiar to spavin is not so pronounced in such cases.

Prognosis.—An animal having hereditary predisposition to spavin is not likely to recover completely whether this predisposition be due to faulty conformation or susceptibility to bone changes. In predicting the outcome, the temperament of the subject is to be taken into account, as well as the character of service the animal is expected to perform. And finally, a very important feature to be noted, is the location of the exostosis. If situated rather high and extending anterior to the hock, there is less likelihood of recovery resulting than where an exostosis is confined to the lower row of tarsal bones. When situated anterior to the tarsus a large exostosis may by mechanical interference to function, cause lameness when all other causes are absent. In making examinations one must not be deceived by the inconspicuous and seemingly insignificant exostosis which has a broad base. In some cases of this kind, dealers style the condition as "rough in the hock" when as a matter of fact, in some instances, incurable spavin lameness develops.

Treatment.—Many incipient cases of spavin yield to vesication and a protracted period of rest. Results depend primarily upon the nature of the affection. However, in every instance if there is involvement of the tibial tarsal (astragalus) bone, complete recovery is highly improbable. When the disease is confined to the lower tarsal bones, lameness subsides as soon as the degenerative changes are checked and ankylosis occurs.

The use of the actual cautery when properly employed constitutes an excellent method of treatment. The "auto-cautery" when equipped with a point of about one-eighth of an inch in diameter and about three-fourths of an inch in length is well suited for this particular operation. Before deciding to cauterize, it is necessary to ascertain the extent of area affected. The nearness of the exostosis to the tibiotarsal articulation can be definitely determined by palpation. The hair over the entire surgical field is clipped and the cautery at white heat is pushed through the overlying soft tissues and into the central part of the exostosis. Care is taken to keep the cautery-point away from the articular margin of the tibial tarsal bone about three-fourths of an inch. No danger will result from cauterizing to a depth of three-fourths of an inch in the average case. Two or three (and not more) centrally located points for penetration with the cautery are sufficient. Experience has shown that several (five or six or more) punctures are not productive of good results. When considerable cicatricial tissue is present, due to the action of depilating vesicants or other chemicals, sloughing of tissue is very apt to follow deep cauterization, if one is not careful to keep the punctures at least one-half inch apart when three are made. It is best, in such cases, to make but two deep penetrations with the cautery but additional superficial punctures may be made if kept about three-fourths of an inch distant and not nearer than this to one another. Sloughing of tissue is not necessarily productive of bad results but there is occasioned an open wound which usually becomes infected and necrosis of tissue may extend into the articulation. No benefit results from sloughing and it should be avoided. In small horses, one deep point of cauterization is sufficient if the osseous tissues are penetrated to a proper depth so that an active inflammation is induced. The cautery may, if necessary, be reintroduced several times. When the field of operation has been properly prepared and it is thought advisable (as where subjects are kept in the hospital for a time), the hock may be covered with cotton and bandaged and no chance for infection will occur.

After cauterization the subject should be kept quiet in a comfortable stall for three weeks; thereafter, if the animal is not too playful, the run of a paddock may be allowed for about ten days and a protracted rest of a month or more at pasture is best. It is unwise in the average case to put an animal in service earlier than two months after having been "fired."

Where cases progress favorably, lameness subsides in about three weeks after cauterization and little if any recurrence of the impediment is manifested thereafter. However, because of violent exercise taken in some instances when subjects are put out after being confined in the stall, a return of lameness occurs and it may remain for several days or in some cases become permanent. No good comes from the use of blistering ointments immediately after cauterization. The actual cautery is a means of producing all necessary inflammation and it should be so employed that sufficient reactionary inflammation succeeds such firing. The use of a vesicating ointment subsequent to cauterization invites infection because of the dust that is retained in contact with the wound. The employment of irritating chemicals in a liquid form following firing is needless and cruel.

In many instances lameness is not relieved and subjects show no improvement at the end of six weeks time and it then becomes a question of whether or not recovery is to be expected even with continued rest and treatment. As a rule, such cases are unfavorable. In one instance the author employed the actual cautery three times during the course of six months and lameness gradually diminished for a year. In this case the spavin was of nearly one year's standing when treatment was instituted. The subject was a nervous and restless but well-formed seven-year-old gelding. Recovery was not complete; recurrent intervals of lameness marked this case, but the horse limped so slightly that the average observer could not detect its existence after the animal had been driven a little way.

Cunean tenotomy has been advocated and practiced by Abildgaard, Lafosse, Peters, Herring, Zuill and others and good results have followed in many cases so treated.

Considering results, the employment of chemicals of various kinds for the purpose of relieving spavin lameness does not compare favorably with firing. Moreover, so many animals have been tortured and needlessly blemished in the attempted cure of spavin that agents which are not of known value, the use of which are likely to result in extensive injury to the tissues, are only to be condemned.

When spavin is bilateral and lameness is likewise affecting both members, prognosis is at once unfavorable. Such cases are often benefited by cauterization but only one leg at a time should be treated.

Bossi's double tarsal neurectomy (division of the anterior and posterior tibial nerves) has undoubtedly been of decided benefit in many cases, but is not at present a popular method of treatment in this country. This operation has its indications, however, and may be recommended in chronic lameness where no extensive exostosis exists which may mechanically interfere with function.

Distension of the Tarsal Joint Capsule. (Bog Spavin.)

Distension of the capsular ligament of the tibial tarsal (tibioastragular) joint with synovia is commonly known as bog spavin. This condition is separate and distinct from that of distension of the sheath of the deep flexor tendon (perforans) though not infrequently the two affections coexist.

Etiology and Occurrence.—Following strains from work in the harness or under the saddle, horses develop an acute synovitis of the hock joint, which often results in chronic synovial distension. Debilitating diseases favor the production of this affection in some animals. It is also frequently observed in young horses and in draught colts of twelve to eighteen months of age. This condition occurs while the subjects are at pasture and often spontaneous recovery results by the time the animals are two years of age.

Fig. 56—Bog spavin. Showing point of view which may be most advantageously taken by the diagnostician in examining for distension of the capsular ligament of the tarsal joint.

Symptomatology.—Bog spavin is recognized by the distended condition of the joint capsule which is prominent just below the internal tibial malleolus and this affection is characterized by a fluctuating swelling which varies considerably in size in different subjects. Except in cases of acute synovitis, lameness is not present and in chronic distension of the capsule of the tarsal joint, no interference with the subject's usefulness occurs. In the majority of instances, the disfigurement which attends bog spavin is the principal objectionable feature. The condition is bilateral in many instances, and in such cases the subjects have a predisposition to this condition or it follows attacks of strangles or other debilitating ailments. Because of a rapid and unusual growth, bilateral affections are of frequent occurrence in some animals.

Treatment.—The most practical method of handling bog spavin consists in aspiration of synovia and injection of tincture of iodin. Discretion should be employed in selecting subjects for treatment, regardless of the manner in which such cases are to be handled. Where there exists chronic distension of the joint capsule of several years' standing in old or weak subjects, needless to say, recovery is not likely to result. When animals are vigorous and two or three months' time is available, treatment may be begun with reasonable hope for success.

The average subject is handled standing and can be restrained with a twitch, sideline and hood. Aspirating needles and all necessary equipment must be in readiness (sterile and wrapped in aseptic cotton or gauze) so that no delay will occur from this cause when the operation has been started. The central or most prominent part of the distended portion of the capsule is chosen for perforation and an area of an inch and a half in diameter is shaved. The skin is cleansed and then painted with tincture of iodin. The sterile aspirating needle is pushed through the tissues and into the capsule with a sudden thrust. With a large and sharp needle (fourteen gauge), synovia can be drawn from the cavity in most instances and the subject usually offers no resistance. By compressing the distended capsule and surrounding structures with the fingers, considerable synovia may be evacuated. In singular instances, no synovia is to be aspirated with the needle, and in such cases the amount of iodin injected needs be increased, possibly twenty-five per cent., as experience will indicate. From two to five cubic centimeters of U.S.P. tincture of iodin is injected through the aspirating needle into the synovial cavity of the joint, and the exterior of the parts are vigorously massaged immediately after injection to stimulate distribution of the iodin throughout the synovial cavity. Where a bilateral affection exists, two or three weeks' time should intervene between the treatments of each leg. A sterile metal syringe equipped with a slip joint for the needle is well adapted to this operation. Lubrication of the plunger with heavy sterile vaseline or glycerin will prevent the syringe from being ruined by the iodin.

Following the injection, the subject is kept in a stall or in a suitable paddock, so that conditions may be observed for four or five days. The object sought by the introduction of iodin is not only for a local effect upon the synovial membranes in checking secretions, but the production of an active inflammation and great swelling, which will remain from four weeks to three months subsequent to the injection. This periarticular swelling should produce and maintain a constant pressure over the entire affected parts for a sufficient length of time until normal tone is re-established.

In some cases, swelling does not develop as the result of a single injection of iodin. When marked swelling has not taken place within five days, none will occur and a repetition of the injection may be made within ten days after the first treatment has been given. One may safely increase the amount of iodin at the second injection in such cases by one-fourth to one-third.

In Europe this method of treating bog spavin has been employed by Leblanc, Abadie, Dupont and others according to Cadiot; but Bouley, Rey, Lafosse and Varrier used it with bad results. Where a perfect technic is executed (and no other is excusable in this operation), no infection will occur if a reasonable amount of iodin is injected. The dilution of iodin with an equal amount of alcohol has been practised by the author in many cases, but later this was found unnecessary.

Other methods of treatment have been used with success. Perhaps the most heroic consists in opening the joint capsule with a bistoury or with the actual cautery. Such practice is too hazardous for general use and is not to be recommended, although good results should follow the employment of such methods if infectious arthritis does not occur.

Line firing over the distended capsule is a practical method of treatment. This is attended with good results in young animals in many cases, but considerable blemish is caused when sufficient irritation is produced to stimulate resolution.

Vesication also is successfully employed in some instances. However, only cases of recent origin in young animals—colts of two years or younger—yield to blistering, and in some affected colts no doubt recovery would have been spontaneous had no treatment been instituted.

Ligation of the saphenous vein at two points, one above and the other below the distended ligamentous capsule, is an old operation, which has undoubtedly given good results in some cases, although it does not seem to be a rational procedure.

After-Care.—After swelling has fully developed—which occurs within a week—the subject is turned to pasture and no attention is necessary thereafter. A gradual subsidence of the swelling occurs and in the average instance, this completely resolves within six or eight weeks.

Complete recovery succeeds the aspiration-and-injection-treatment in about seventy-five per cent of cases as the result of one operation, and subjects may be gradually and carefully returned to work in about sixty days after treatment has been given.

Distension of the Tarsal Sheath of the Deep Digital Flexor.
(Thoroughpin.)

The terms "thoroughpin" or "throughpin" are translations from the French *vessignon chevillé* and have the same significance. They are so named because of the diametrically opposed distensions of the sheath of the deep flexor tendon in such manner that the distensions appear to be due to a supporting peg.

Anatomy.—The theca through which the deep digital flexor (perforans) plays in the tarsal region, begins about three inches above the inner tibial malleolus and extends about one-fourth of the way down the metatarsus. The posterior part of the capsular ligament of the hock joint is very thick in its most dependent portions and is in part cartilaginous, forming a suitable groove for the passage of the deep flexor tendon.

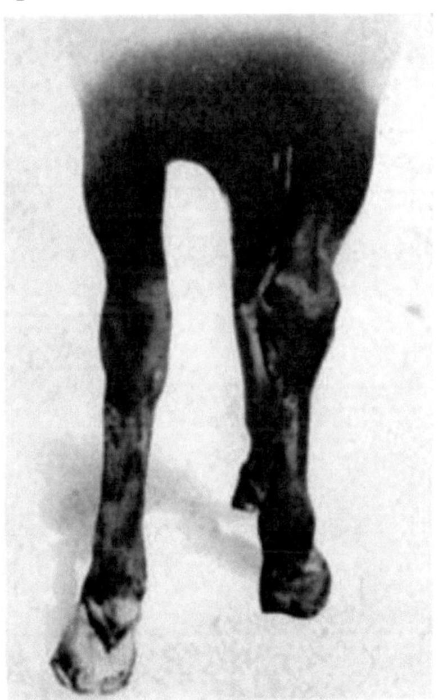

Fig. 57—Thoroughpin. Showing distension of the sheath of the deep flexor tendon as it protrudes antero-externally to the fibular tarsal bone (calcaneum).

Etiology and Occurrence.—Strains and sequellae to debilitating diseases constitute the usual causes of this affection. As a result of acute synovitis a chronic synovial distension of the tarsal sheath occurs. Bog spavin is often present in case of thoroughpin but the two conditions are separate and distinct excepting in that both may occur simultaneously and as the result of the same cause. Some animals are undoubtedly predisposed to disease of synovial structures. The average horse that has been subjected to hard service on pavements or hard roads at fast work suffers synovial distension of bursae, thecae or of joint capsules. Some of the well bred types such as the thoroughbred horses may be subjected to years of hard service and still remain "clean limbed" and free from all blemishes. Thus it seems that subjects of rather faulty conformation, animals having lymphatic temperaments and the coarse-bred types, are prone to synovial disturbances such as thoroughpin, bog spavin, etc., sometimes having both legs affected.

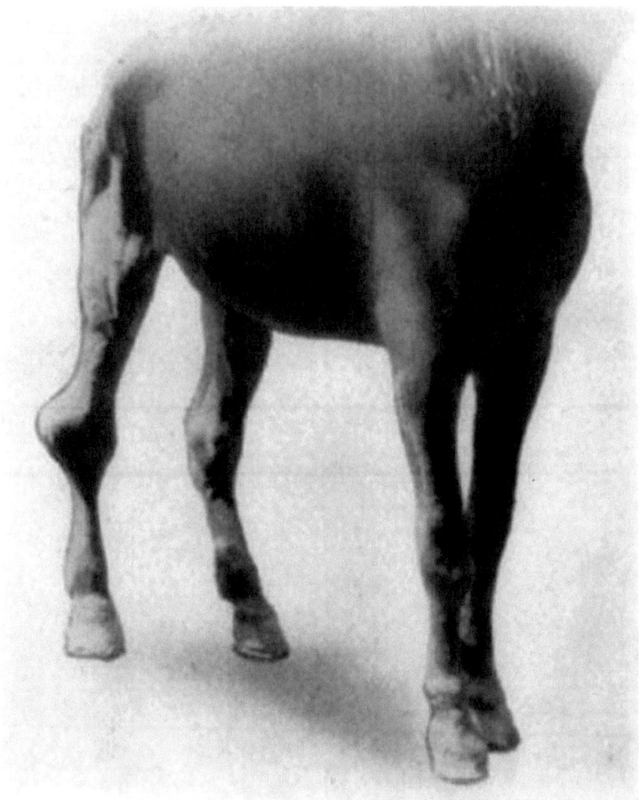

Fig. 58—Fibrosity of tarsus as a complication in chronic thoroughpin.

Symptomatology.—Thoroughpin is characterized by a distended
condition of the tarsal sheath which is manifested by protrusions
anterior to the tendo Achillis. However, where but moderate distension of the sheath exists, there is little, if any, bulging on the mesial
side of the hock and but a small hemispherical enlargement is presented on the outer side of the tarsus, anterior to the summit of the
os calcis. In some instances the protruding parts assume large proportions, but always, because of the relationship between the fibular
tarsal bone (calcaneum) and the tendon sheath, the larger protrusion is situated mesially.

During the acute inflammatory stage there is marked lameness present but this soon subsides when local antiphlogistic agents are applied to the parts. In fact, spontaneous relief from lameness usually results in the course of ten days' time following the appearance of thoroughpin. No lameness marks the advent of this affection when it develops as the result of continuous strain and concussion occasioned by hard service, and local changes tend to remain in *status quo.*

Fig. 59 — Another view of same case as illustrated in Fig. 58.

Treatment. — Rest and the local application of heat or cold will suffice to promote resolution of acute inflammation and lameness when present will subside within two weeks. In chronic affections, however, the matter and manner of effecting a correction of the condition—distended tarsal sheath—merit careful consideration. While drainage of distended thecae and bursae by means of openings made with hot irons was practiced by the Arabs, centuries ago, and good results have attended such heroic corrective measures, nevertheless the occasional serious complications which result from infection likely to be introduced in following such procedures, cause the prudent and skilful practitioner to employ safer methods of treatment.

The application of blistering agents is of no value in stimulating resorption of an excessive amount of synovia in chronic cases and the actual cautery when employed without perforation of the synovial structure, is of little benefit. Trusses or mechanical appliances for the purpose of maintaining pressure upon the distended parts are of no practical value because of the great difficulty of keeping such contrivances in position. They usually cause so much discomfort to the subject that they are not tolerated.

A very practical and fairly successful method of treatment consists in the aspiration of a quantity of synovia and injecting tincture of iodin. Cadiot recommends the drainage of synovia with a suitable trocar and cannula and injecting a mixture consisting of tincture of iodin, one part, to two parts of sterile water, to which is added a small quantity of potassium iodid. The latter agent is added to prevent precipitation of the iodin. This authority (Cadiot) further advocates the removal of practically all of the synovia that will run out through the cannula and the immediate introduction of as much as one hundred cubic centimeters of the above mentioned iodin solution. This solution is allowed to remain in the synovial cavity a few minutes and by compressing the tissues surrounding the tendon sheath, the evacuation of as much of the contents of the synovial cavity as is practicable, is effected. Subsequently the subject is allowed absolute rest and more or less inflammatory reaction follows. In some cases there occur marked lameness and some febrile disturbance, but where a good technic is carried out, no bad results follow. At the end of four weeks' time, horses so treated may be returned to service, but the full beneficial effect of such treatment is not experienced until several months' time have elapsed.

Where good facilities for executing a careful technic in every detail are at hand, incision of the tarsal sheath, evacuation of its contents and uniting its walls again by means of sutures and providing for drainage with a suitable drainage tube, may be practiced. This manner of treatment has been satisfactory in the hands of a number of surgeons.

Capped Hock.

Enlargements which occur upon the summit of the os calcis, whether hypertrophy of the skin and subcuticular fascia, the result of injury or repeated vesication, distension of the subcutaneous bursa or injury to the superficial flexor tendon (perforatus) or its sheath, are generally known as capped hock. However, the term should be restricted to use in reference to distensions of synovial structures of that region.

Etiology and Occurrence.—Usually there occurs a hygromatous involvement of the subcutaneous bursa due to contusion. As in bog spavin, following certain infectious diseases (influenza, purpura hemorrhagica, etc.) there remains a distended condition of the subcutaneous bursa, after swelling of the member has subsided. In feeding pens where numbers of young mules are kept in crowded quarters many cases may be observed. In some instances where violent contusions result from kicking cross-bars of wagon shafts (by nymphomaniacs or in habitual kickers where there is opportunity for doing such injury) the superficial flexor tendon and its synovial apparatus are injured and a more serious condition may result.

Symptomatology.—In acute and extensive inflammation of the parts, lameness is present, but in the average case no inconvenience to the subject results. The prominent site of the affection is cause for an unsightly blemish. This is undesirable, particularly in light-harness or saddle horses. These affections are characterized by a fluctuating mass which has a thin wall and in all cases of long standing the condition is painless.

By careful palpation one may readily distinguish between a hygromatous condition of the superficial bursa and involvement of the underlying structures. Affection of the expanded portion of the flexor tendon and contiguous structures makes for an organized mass of tissue which is somewhat dense and in some instances painful to the subject when manipulated. This is particularly noticeable in cases where the parts are regularly and repeatedly injured as in habitual kickers.

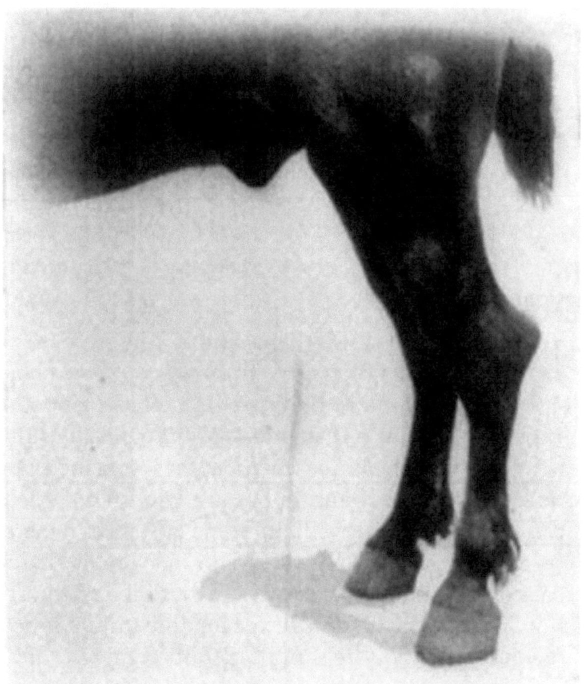

Fig. 60—"Capped hock." Distension of the bursa over the summit of the os calcis.

Treatment.—In acute inflammation, antiphlogistic applications are indicated and the subject must be kept quiet. The matter of bandaging the hock is a difficult problem in some cases and needs be done with care. As has been previously stated in this volume, the tarsus needs to be well padded with cotton before the bandages are applied and only a moderate degree of tension is employed in applying the bandages lest anemic-necrosis result from pressure. In distension of the superficial bursa, after clipping the hair over a liberal area and preparing the skin by thoroughly cleansing and painting with tincture of iodin, the capsule is incised with a bistoury. An incision about an inch in length, situated low enough to provide drainage, is made through the tissues and the contents are evacuated. Tincture of iodin is injected into the cavity and the parts are covered with cotton and bandaged. No after-care is necessary except to retain the dressing in position, which is not difficult in the average case if the subject is kept tied. If much resistance is exhibited, such as extreme flexion of the bandaged hock, the animal may be put in a sling and little if any objection to the bandage will be offered thereafter. The wound may be dressed at the end of forty-eight hours and no redressing will be necessary in the average instance if infection is not present. But slight local disturbance and little distress to the subject result in cases so treated even when infection occurs, but a good technic is possible of execution in most instances and no infection should take place.

The surgical wound heals in two or three weeks and inflammation gradually subsides. Bandages are retained one or two weeks, as the case may require, and subsequently a good wound lotion may be employed several times daily. A good lotion for such cases as well as in many others has long been employed with success by Dr. A. Trickett of Kansas City. It consists of approximately equal parts of glycerin, alcohol and distilled extract of witch hazel, to which is added liquor cresolis compositus, two percent, and coloring matter q.s.

Complete resolution does not occur in the average case. There remains some hyperplastic tissue and even where the enlargement is slight, the prominent situation of the affection precludes its being unnoticed.

In disease of the flexor tendon and its bursa where contiguous inflammation of tissue is present, the parts are blistered or fired. Line firing is beneficial in such instances but in all cases the cause is to be removed if possible.

Rupture and Division of the Long Digital Extensor (Extensor Pedis).

Etiology and Occurrence.—Because of the fact that the long digital extensor is the only extensor of the phalanges of the pelvic limb, its rupture or division constitutes a troublesome condition, which in some cases does not readily respond to treatment.

Rupture of this tendon may occur during work on rough and uneven roads, particularly in range horses that are ridden over ground that is burrowed by gophers or prairie dogs; in such cases, horses are apt to suddenly and violently turn the foot in position of volar flexion, thereby causing undue strain to the digital extensor and its rupture sometimes follows. In foals of one or two days of age, this tendon is sometimes found parted or ruptured and the condition may be bilateral.

As the result of accidents, the digital extensor may be divided and when the wound becomes contaminated, as it does because of the marked volar flexion (knuckling) which occurs during the course of this affection, regeneration of tissue is checked and recovery is tardy.

Symptomatology.—There is no interference with ability to sustain weight in such cases, when the foot is placed in normal position; but immediately upon attempting to walk, the toe is dragged, and if weight is borne with the affected member, it comes upon the anterior face of the fetlock. The flexors are not antagonized and if there be an open wound the parts soon become contaminated; or, in rupture, if animals travel about very much, there soon occurs necrosis of the tissues of the anterior fetlock region and the condition is rendered incurable. Cases are reported of animals that have suffered rupture of the long digital extensor and the subjects learned to throw the member forward during extension, substituting for the extensor tendon the pendulum-like momentum which the foot affords when so employed; and a walking and even a trotting pace was possible without doing injury to the fetlock region.

Where a subcutaneous division exists as in rupture, the divided ends of the tendon may be definitely recognized by palpation.

Treatment.—Subjects are best put in slings and kept so confined until regeneration of tendinous structures has been completed. This requires from six weeks to two months' time. In addition, the extremity is kept in a state of extension by means of suitable splints and shoes,—a shoe equipped with an extension at the toe and perforated so that a steel brace may be hooked into the perforation and the brace fashioned to be buckled to the upper metatarsal region. When braces are placed in front of the foot, great care is necessary in properly padding the member with cotton lest sloughing from pressure occurs at the coronet; but this does not apply in rupture of extensors so much as where flexors are ruptured.

Open wounds are treated along general surgical lines, dressed as frequently as occasion demands, and recovery will be complete in a few months' time unless much of the tendon has been destroyed. In one instance, the author had occasion to observe such a condition, which, because of the extensive destruction of tendon and lack of facilities for giving proper attention to the subject, results were so unfavorable that it was deemed necessary to destroy the animal.

Wounds From Interfering.

When, during locomotion, injury is inflicted upon the mesial side of an extremity by the swinging foot of the other member, the condition is termed interfering.

Etiology and Occurrence.—Faulty conformation, bad shoeing and over-work are the principal causes of interfering. Horses that are "base narrow" or that have crooked legs are quite apt to interfere. Shoes that are put on a foot that is not level or applied in a twisted position, or shoes wide at the heel will often cause interfering and injury. Animals that are driven at fast work until they become nearly exhausted may be expected to interfere. Such cases are frequently observed in young horses that are driven over rough roads, particularly when so nearly exhausted or weakened from disease or inanition that the feet are dragged forward rather than picked up and advanced in the normal manner.

Symptomatology.—Wounds inflicted by striking the extremities in this manner present various appearances and occasion dissimilar manifestations. The hind legs are almost as frequently affected as the front and the fetlock region is most often injured, though wounds may be inflicted to the coronet. In front, the carpus is sometimes the site of injury.

When only an abrasion is caused, little if any lameness occurs, but where interfering is continued and nerves are involved or subfascial infection and extensive inflammation succeed such abrasions, marked lameness and evidence of great pain are manifested. Frequently, in chronic cases affecting the hind leg, the fetlock assumes large proportions, and at times during the course of every drive the subject strikes the inflamed part, immediately flexing and abducting the injured member, and the victim hops on the other leg until pain has somewhat subsided.

Interfering is much more serious in animals that are used at fast work than in draft horses. In light-harness or saddle horses, it may render the subject practically valueless or unserviceable if the condition cannot be corrected.

Treatment. — Wherever possible, cause is to be removed and if animals are properly used, ordinary interfering wounds will yield to treatment. If the shoeing is faulty, this should be corrected, the foot properly prepared and leveled before being shod and suitable shoes applied. In young animals that become "leg-weary" from constant overwork, rest and recuperation are necessary to enhance recovery. In such cases it will be found that very light shoes, frequently reset, will tend to prevent injury to the fetlock region such as characterizes these injuries of hind legs.

Palliative measures of various kinds are employed where cause is not to be removed and a degree of success attends such effort. In draft horses or animals that are used at a slow pace, shields of various kinds are strapped to the extremity and protection is thus afforded. Or, large encircling pads of leather, variously constructed, serve to cause the subject to walk with the extremities apart.

Interfering shoes of different types are of material benefit in many instances. Often the principle upon which corrective shoeing is based is that the mesial (inner) side of the foot is too low; the foot is consequently leveled and the inner branch of the shoe is made thicker than the outer, altering the position of the foot in this way. This is productive of desirable results. However, much depends upon the manner in which the foot in motion strikes the weight-bearing member as to the corrective measures that are indicated. This belongs to the domain of pathological shoeing and the reader is referred to works on this subject for further study of this phase of lameness.

Lymphangitis.

Excluding glanders, in the majority of instances, lymphangitis in the horse, such as frequently affects the hind legs, is due to the local introduction of infectious material into the tissues as a result of wounds. However, one may observe in some instances an acute lymphangitis which affects the pelvic limbs of horses and no evidence of infection exists. Consequently, lymphangitis may be considered as *infectious* and *non-infectious*.

INFECTIOUS LYMPHANGITIS.

Etiology and Occurrence.—Traumatisms of the legs frequently result in infection and when such injuries are near lymph glands, even though the degree of infection be slight, more or less disturbance of function of the muscles in the vicinity of such glands occurs and lameness follows.

The prescapular, axillary and cubital lymph glands when in a state of inflammation, cause lameness of the front leg, and the superficial inguinal and deep inguinal lymph glands not infrequently become involved also. Because of the location of these lymph glands, they are subject to comparatively frequent injury and inflammation, causing lameness more often than other lymph-gland-affections.

Small puncture wounds in the region of the elbow are often met with. These may be inflicted when horses lie down upon sharp stumps of vegetation or shoe-calk injuries may be the means of introducing contagium, and an infectious inflammation results. Abscess formation, the result of strangles or other infection in the prescapular glands, may be observed at times. Following castration, the inguinal lymph glands may become involved in an infectious inflammation and locomotion is impeded to a marked degree. Horses running at pasture sometimes become injured by trampling upon pieces of wood, causing one end of these or of various implements to become embedded in the soft earth and the other end to enter at the inguinal region and even penetrate the tissues to and through the skin and fascia just below the perineal region.

Nail punctures resulting in infection frequently cause an infectious lymphangitis and a marked and painful swelling of the legs supervenes.

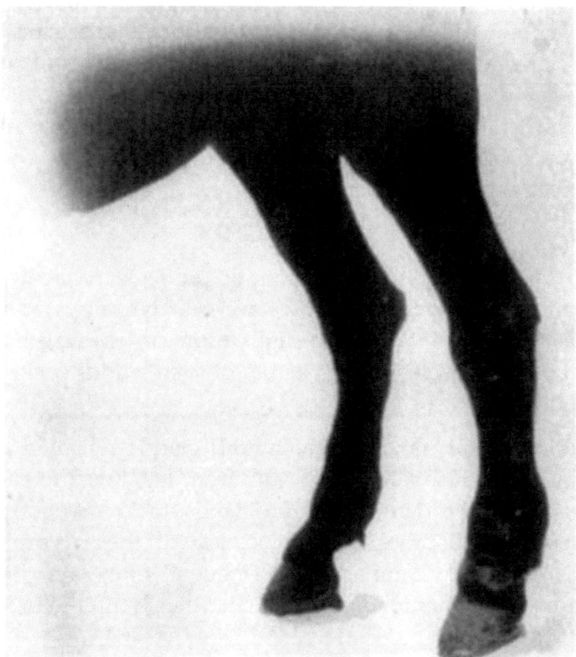

Fig. 61—Chronic lymphangitis. Showing hypertrophy of the left hind leg, due to repeated inflammation.

Symptomatology.—Lameness, mixed or swinging-leg, signalizes the presence of acute lymphangitis. There is always more or less swelling present and manipulation of the affected parts gives pain to the subject. Depending upon the character of the infection and its extent, there is presented a varying degree of constitutional disturbance. There may be a rise in temperature of from two to five degrees, and in such instances there is an accelerated pulse. Where much intoxication is present, anorexia and dipsosis are to be noticed.

Swelling may increase gradually and in time discharge of pus may take place spontaneously without drainage being provided for, if the character of the infection does not cause early death. In these

360

cases lameness is pronounced and the cause of the disturbance is to be sought, particularly if the condition be due to a nail puncture.

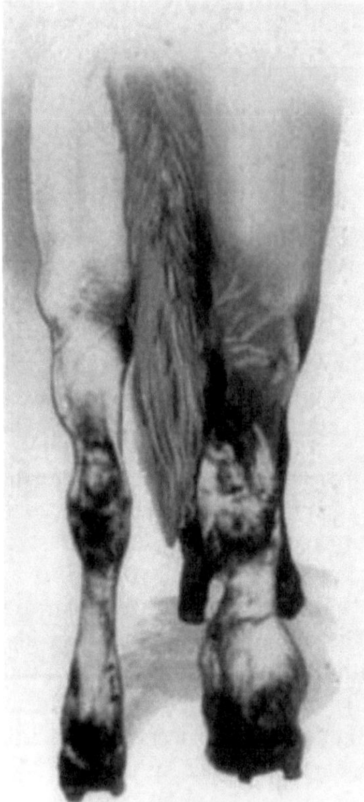

Fig. 62—Elephantiasis.

Treatment.—Location of the site of injury is advisable in all cases and in some instances provision for drainage, as in puncture wounds, is helpful. Locally, curettage and the application of suitable antiseptics are indicated. Hot fomentations are beneficial and should be continued for several days if necessary, to stimulate resolution. A brisk purge should be admintered at the onset and strychnin, because of its indirect stimulative effect upon the circulation together with its tonic effect upon the musculature, is beneficial.

In all such cases rational treatment, good hygiene and careful nursing are the principal factors which stimulate recovery. Individual resistance or lowered vitality has a marked influence on the course of this affection.

NON-INFECTIOUS LYMPHANGITIS.

This type of lymphangitis is associated with, or the result of, a derangement of digestion. It affects heavy draft horses, rarely other types of animals, and involves one or both hind legs.

Occurrence.—In healthy and well nourished horses irregularly used, this affection may suddenly manifest itself. It occurs in singular instances in mares that are in advanced pregnancy even when such animals are at pasture. Usually, however, this malady is found in heavy draft horses that have been kept stabled from one to three days.

Symptomatology.—At the outset in severe cases, there is elevation of temperature, labored breathing, accelerated pulse, anorexia and more or less swelling of the affected members. Swelling is very painful and when the affected legs are palpated, pain is manifested by flinching. The inguinal lymph glands are often swollen but in some cases they are not affected in any perceptible degree. In the average case suppuration does not occur and when conditions are favorable, resolution is complete within ten days. The extent of the involvement and the intensity of the affection vary materially in different cases and a chronic lymphangitis may succeed the acute attacks and finally in some instances, elephantiasis results.

Treatment.—An active purgative should be given at once and in the ordinary case, stimulants are indicated. If marked distress is present, morphin is given and where there is much rise of temperature, cold drinking water is offered in abundance and catharsis is enhanced by enemata. Locally, hot applications are of benefit. Hot towels or cotton held in position by bandages and kept soaked with warm water will relieve pain and stimulate resolution. Diuretics may be of benefit and anodyne applications are to be employed with profit in some cases. Walking exercise, if not indulged in to excess, is helpful as soon as acute inflammation has subsided. By giving careful attention to the regimen and providing regular exer-

cise for susceptible subjects, this type of lymphangitis is often fore-
stalled.

FOOTNOTES:

[34] Manual of Veterinary Physiology. Page 610.

[35] Manual of Veterinary Physiology, page 601.

[36] Case report at meeting of the Iowa State Veterinary Medical Association, Jan., 1904, by Dr. S.H. Bauman, Birmingham, Ia.

[37] Regional Veterinary Surgery and Operative Technique, by John A.W. Dollar, M.R.C.V.S., F.R.S.E., M.R.I., page 733.

[38] As quoted by A. Liautard, M.D., V.M., American Veterinary Review, Vol. 37, page 667.

[39] Quoted by Prof. Liautard, American Veterinary Review, Vol. 33, page 190.

[40] Traite de Thérapeutique Chirurgical des Animaux Domestique par P.J. Cadiot et J. Almy, Tome second, page 460.

[41] Traite de Thérapeutique Chirurgical, Tome second, page 465.

[42] Luxation of the Femur, by Wm. V. Lusk, Veterinary Surgeon, U.S. Cavalry, American Veterinary Review, Vol. 21, page 254.

[43] Because of the intimacy of the psoas major (p. magnus) and the iliacus they are sometimes called iliopsoas.

[44] Dr. John Scott, Peoria, Ill., in The American Veterinary Review, Vol. 16, page 16.

[45] Annotation on Surgical Items, by Drs. L.A. and Edward Merillat, American Veterinary Review, Vol. 31, page 358.

[46] W.L. Williams in American Veterinary Review, Vol. 21, page 452.

[47] Geo. H. Berns, D.V.S., report, American Veterinary Medical Association, 1912, page 238.

[48] Joseph Hughes, M.R.C.V.S., in the Chicago Veterinary College Quarterly Bulletin, Vol. 10, page 15.

[49] Traite de Therap. Chir. Cadiot et Almy, Tome second, page 480.

[50] E. Wallis Hoare, F.R.C.V.S., American Veterinary Review, Vol. 27, page 1189.

[51] Discussions on paper entitled "The Spavin Group of Lamenesses," by W.L. Williams, Carl W. Fisher and D.H. Udall, Proceedings of American Veterinary Medical Association, 1905.

[52] "Hock-Joint Lameness," by Dr. James McDonough, Proceedings of the A.V.M.A., 1913, page 545.